England's Coastal Heritage

England's Coastal Heritage

A review of progress since 1997

Peter Murphy

ENGLISH HERITAGE

Published by English Heritage, The Engine House, Fire Fly Avenue, Swindon SN2 2EH
www.english-heritage.org.uk
English Heritage is the Government's lead body for the historic environment.

Images © Peter Murphy (except as otherwise shown)

First published 2014

ISBN 978 1 84802 107 5

Product code 51696

British Library Cataloguing in Publication data
A CIP catalogue record for this book is available from the British Library.

For more information about images from the English Heritage Archive, contact Archives Services Team, The Engine House, Fire Fly Avenue, Swindon SN2 2EH; telephone (01793) 414600.

Brought to publication by Victoria Trainor, Publishing, English Heritage.

Typeset in Charter 9.5pt on 11.75pt
Edited by Val Kinsler, 100% Proof
Indexed by Ann Hudson
Page layout by Andrea Rollinson at Ledgard Jepson Ltd

Printed in the UK by Butler Tanner and Dennis Ltd

CONTENTS

PREFACE

In 1993 English Heritage (EH) commissioned Professors Michael Fulford (University of Reading) and Timothy Champion (University of Southampton) and Dr Antony Long (then University of Southampton, now a Professor at Durham) to undertake a review of coastal archaeology in England, to assess what was known and to develop an agenda for the future. The book published in 1997, *England's Coastal Heritage*, (Fulford *et al* 1997) was a milestone in English coastal archaeological research, including major contributions from other leading British archaeologists and sea-level researchers. It would now be described as an Archaeological Research Framework, anticipating modern frameworks, which all include the components of Resource Assessment, Research Agenda and Research Strategy.

One of the main recommendations was for a new, comprehensive national coastal survey, which led to initiation of the English Heritage Rapid Coastal Zone Assessment Survey (RCZAS) programme in the late 1990s, under the direction of Steve Trow. From 2005 the present writer assumed responsibility for the programme. It was termed *rapid* and it has been, so far as available funding has permitted, but has not been *quite* as rapid as originally hoped. It is still continuing, fifteen years on from 1997. It has been running for long enough to have its own history and, like other national survey programmes, for example the British Geological Survey (BGS), originating in 1837 (Flett 1937), has undergone changes in objectives and emphasis since its beginning. Technical and technological development, regional variations and changes of emphasis have led to repeated revision of the Briefs that have guided it. The current emphasis in EH on designation, and thereby protection, was not considered in the earliest surveys. The end-of-project reports produced have undoubtedly improved through time in terms of their comprehensiveness and applicability, so that a 2012 report is likely to be more useful now than those of the late 1990s. Some of the earlier reports were specifically archaeological, whereas later reports covered a wider range of historic assets. A change of emphasis is inevitable with any long-running programme (http://www.english-heritage.org.uk/professional/advice/advice-by-topic/marine-planning/shoreline-management-plans/rczas-reports/).

This publication refers primarily to studies of the coastal historic environment, many of which (though by no means all), were funded and developed by EH. It includes a review of the national RCZAS programme, but also refers extensively to other regional work. The RCZAS is now embedded in EH's National Heritage Protection Plan (NHPP: Activity 3A2, Coastal Survey), but at the time of writing the surveys are incomplete. Phase 1 (Desk-Based Assessment) surveys for the South-West are only now starting, and Phase 2 (Field Survey) has not been initiated for the South-East. So, the reader might reasonably ask: why a review *now*? Although deferring this publication until completion of the RCZAS in the next NHPP round after 2014 would ideally be desirable, there are sound reasons for a summary and discussion at this point. First, future funding for the RCZAS is not assured in the longer term – we do not know how funding for the historic environment might change, given the country's present economic situation. Secondly, I initiated many of the surveys and have been involved in all of them, so far, and I expect to be retired before final completion of the programme. This gives some insight and experience that ought to be set down in text now.

At first I thought that the original format and contents structure of the original book would be adhered to very closely in this review. However, that has become less possible as writing has proceeded. Some issues that seemed significant in 1997 no longer retain their importance. Conversely, other aspects that were not then considered at all have attained greater prominence – especially the vastly increased understanding of submerged prehistoric archaeology, landscapes and coastal change, derived largely from projects undertaken with funding from the Aggregates Levy Sustainability Fund (ALSF). In addition, work by the British Museum (Natural History), British Academy-funded, Ancient Human Occupation of Britain (AHOB) project has extended enormously the time scale of a human presence, back to almost a million years ago. Fulford and his colleagues explicitly excluded 20th century structures, including those of both World Wars, but more recent survey has demonstrated their enormous significance in defining the coastal historic environment. From a Flood and Coastal Erosion Risk Management (FCERM) perspective, there has been much greater involvement of English Heritage with Defra (Department for Environment, Food and Rural Affairs) and Environment Agency (EA) initiatives. Above all, the perception that climate change will impact significantly, and mostly adversely, on the coastal historic environment gives a special urgency to a new review.

For simplicity many of the bibliographic references used here refer to the RCZAS reports themselves, even where information within them is derived from earlier sources, and so the RCZAS reports are secondary. This is inevitable if the bibliography is to be restrained within reasonable limits. Readers wishing to inspect the original sources are referred to the references cited in the RCZAS reports.

Peter Murphy

ACKNOWLEDGEMENTS

I am extremely grateful to Robin Taylor who first contacted the (then) Maritime Archaeology Team to suggest publications arising from current work. I would like to thank Martin Bell for reading and commenting on the manuscript, Val Kinsler for her copy editing work and Victoria Trainor for bringing this book to publication.

I am also most grateful to all those who have allowed me to reproduce images throughout this book, especially the late Ron Hall, who donated his images 'for the greater good', Ellen Heppell from the Field Archaeology Unit, Essex County Council, Simon Parfitt from the Ancient Human Occupation of Britain Project, Natural History Museum, Nick Powe from Kents Cavern, and Ian Bell and John Vallender for their redrawing of draft material.

Other images were kindly provided by Archaeological Research Services Ltd; DP World London Gateway and Oxford Archaeology; Historic Environment Service; NPS Archaeology; Norfolk County Council; Norfolk Museums Service; SeaZone Solutions Limited; Suffolk County Council Archaeological Service; Sussex Past; The Hampshire and Wight Trust for Maritime Archaeology; The Visual and Spatial Technology Centre, University of Birmingham; United Kingdom Hydrographic Office, Maritime and Coastguard Agency; and Wessex Archaeology.

SUMMARY

The publication of *England's Coastal Heritage*, (Fulford *et al* 1997), arising from English Heritage's first strategic initiative into coastal archaeology published in 1997, presented a milestone in research; but it also recommended a new comprehensive survey of the coast and set out a range of research questions and management approaches. To a large extent subsequent studies, including EH's Rapid Coastal Zone Assessment surveys, but also many other more specific surveys and excavations, often funded by EH but not always, have met Fulford and his colleagues' recommendations. Work on the Palaeolithic of the modern coastal zone and surveys of offshore, formerly terrestrial, sites has involved work that was not envisaged in the 1990s.

Writing in 2013, around 15 years on, there is now a need to review what has been achieved and what remains to be done.

The present book provides an introduction to the historic environment of the coast, sets out a summary of recent methodological advances, considers what we know now about coastal change and the meaning of coastal historic assets on it. It also re-considers research questions for the future and the future management. Above all the final imperative is to adapt the coastal historic environment and its study to the future impacts of climate change, and at a time of wider economic difficulty, and to determine the best way to do so.

LIST OF ABBREVIATIONS

Abbreviations are explained the first time that they are used in the text, and will be used without explanation thereafter.

AA   Anti Aircraft
AHOB   Ancient Human Occupation of Britain Project
ALSF   Aggregates Levy Sustainability Fund
AONB   Area of Outstanding Natural Beauty
BGS   British Geological Survey
BP   Before Present, AD 1950 (for 14C dates).
            Cal BC or cal AD are used for calibrated
            radiocarbon dates.
CCRA   Climate Change Risk Assessment
CERA   English Heritage Coastal Estate Risk Assessment
DCMS   Department for Culture, Media and Sport
Defra   Department for Environment, Food and
            Rural Affairs
EA   Environment Agency
EH   English Heritage
FCERM   Flood and Coastal Erosion Risk Management
GIS   Geographic Information System
GPS   Global Position System
HER   Historic Environment Records (maintained by
            Local Authorities)
IFCA   Inshore Fisheries and Conservation Authority
LiDAR   Light Detection and Ranging
MIS   Marine Isotope Stage
NAP   National Adaptation Programme (for Climate
            Change)

NHPP   National Heritage Protection Plan
NMP   National Mapping Programme
            (of aerial surveys)
NMR   National Monuments Record, predecessor
            of the NRHE
NPPF   National Planning Policy Framework
NRHE   National Record of the Historic Environment
            (maintained by EH)
OS   Ordnance Survey
OSL   Optically Stimulated Luminescence
PLUTO   Pipe Line Under the Ocean
RARF   Regional Archaeological Research Framework
RCZAS   Rapid Coastal Zone Assessment Survey
REC   Regional Environmental Characterisation
ROC   Royal Observer Corps
RSL   Relative Sea Level
SLI   Sea Level Index point, a radiocarbon date
            thought to give a reliable indication of sea level
            at a given time.
SMP   Shoreline Management Plan
SMR   Sites and Monuments Record
SPA   Special Protection Area
SSSI   Site of Special Scientific Interest
UKCP09   United Kingdom Climate Projections 2009
UKHO   United Kingdom Hydrographic Office
WW I   First World War
WW II   Second World War

THE RCZAS REPORTS AND REFERENCING

Rapid Coastal Zone Assessment Survey (RCZAS) reports are listed below in clockwise order around the English coast, beginning in the North East. There will be repeated references to these reports in the text so, to avoid cluttering it with full references, *abbreviations* will be used, as below; for example, NE1, 23 = Tolan-Smith 2008, p 23. Data were incorporated into this text from the RCZAS reports up until 31st January 2014.

Records of historic environment assets in these reports have been taken directly from Local Authority Historic Environment Records (HERs) or the National Record of the Historic Environment (NRHE), or else are new data obtained during the surveys. In the latter case the RCZAS reports themselves are the primary published sources. Otherwise, especially in Phase 1 reports, data are taken from previously published sources. It has proved necessary to be judicious in referring to primary sources to limit the length of the bibliography in this review volume. Full references to original sources will be found in the RCZAS reports themselves.

Please note that the references given in this section are *not* duplicated in the bibliography.

North-East

NE1 Tolan-Smith, C 2008 *North East Rapid Coastal Zone Assessment (NERCZA)*. Archaeological Research Services Report **2008/81**. Gateshead: ARS
http://www.english-heritage.org.uk/publications/nercza-final-report/

NE2 Johnson, B 2009 *North East Rapid Coastal Zone Assessment (NERCZA). Executive Summary*. Archaeological Research Services Report **2009/22**. Gateshead: ARS
http://www.english-heritage.org.uk/publications/nercza-executive-summary/

NE3 Waddington, C 2010 *Low Hauxley, Northumberland: a review of archaeological interventions and site condition*. Archaeological Research Services Report **2010/25**. Gateshead: ARS
http://www.english-heritage.org.uk/publications/low-hauxley-review-archaeological-work/

NE4 Bacilieri, C, Knight, D and Radford, S 2008 *North East Rapid Coastal Zone Assessment Survey. Air Survey Mapping Report. English Heritage National Mapping Programme. HEEP Project Number: 3929*. Swindon: English Heritage
http://www.english-heritage.org.uk/publications/nercza-aerial-survey/

NE5 Burn, A 2010 *North East Rapid Coastal Zone Assessment: Phase 2*. Archaeological Research Services Report **2010/42**. Gateshead: ARS.
http://www.english-heritage.org.uk/publications/nercza-phase2/

NE6 Johnson, B and Waddington, C 2011 *Brief Statement on Rescue Recording of an Eroding Inter-tidal Peat Bed Containing Prehistoric Worked timber and Human and Animal Footprints*. Gateshead: ARS
http://www.english-heritage.org.uk/publications/rescue-recording-eroding-inter-tidal-peat-bed/

Yorkshire and Lincolnshire

YL1 Buglass, J and Brigham, T 2008a *Rapid Coastal Zone Assessment. East Riding of Yorkshire. Whitby to Reighton*. English Heritage Project **3729**. Humber Archaeology Report No. **238**. Kingston-upon-Hull: Humber Archaeology.
http://www.english-heritage.org.uk/publications/rczas-whitby-to-reighton/

YL2 Brigham, T, Buglass, J and George, R 2008 *Rapid Coastal Zone Assessment. Yorkshire and Lincolnshire. Bempton to Donna Nook*. English Heritage Project **3729**. Humber Archaeology Report No. **235**. Kingston-upon-Hull: Humber Archaeology
http://www.english-heritage.org.uk/publications/rczas-bempton-donna-nook/

YL3 Buglass, J and Brigham, T 2008b *Rapid Coastal Zone Assessment. Yorkshire and Lincolnshire. Donna Nook to Gibraltar Point*. English Heritage Project **3729**. Humber Archaeology Report No. **238**. Kingston-upon-Hull: Humber Archaeology
http://www.english-heritage.org.uk/publications/rczas-donna-nook-gibraltar-point/

YL4 Buglass, J and Brigham, T 2007 *Rapid Coastal Zone Assessment. Yorkshire and Lincolnshire. Gibraltar Point to Sutton Bridge*. English Heritage Project **3729**. Humber Archaeology Report No. **237**. Kingston-upon-Hull: Humber Archaeology
http://www.english-heritage.org.uk/publications/rczas-gibraltar-point-norfolk/

YL5 Buglass, J and Brigham, T 2011 *Rapid Coastal Zone Assessment Survey. Yorkshire and Lincolnshire. Whitby to Reighton. Phase 2*. English Heritage Project **3729**. Humber Archaeology Report No. **327**. Kingston-upon-Hull: Humber Archaeology
http://www.english-heritage.org.uk/publications/rczas-whitby-reighton-phase2/

YL6 Brigham, T and Jobling, D 2011 *Rapid Coastal Zone Assessment Survey. Yorkshire and Lincolnshire. Bempton to Donna Nook. Phase 2*. English Heritage Project **3729**. Humber Archaeology Report No. **324**. Kingston-upon-Hull: Humber Archaeology
http://www.english-heritage.org.uk/publications/rczas-bempton-donna-nook-phase2/

YL7 Jobling, D and Brigham, T 2011 *Rapid Coastal Zone Assessment Survey. Yorkshire and Lincolnshire. Donna Nook to Gibraltar Point. Phase 2*. English Heritage Project **3729**. Humber Archaeology Report No. **325**. Kingston-upon-Hull: Humber Archaeology
http://www.english-heritage.org.uk/publications/rczas-gibraltar-point-norfolk-phase2/

YL8 Jobling, D and Brigham, T 2011 *Rapid Coastal Zone Assessment Survey. Yorkshire and Lincolnshire. Gibraltar Point to Sutton Bridge. Phase 2*. English Heritage Project **3729**. Humber Archaeology Report No. **326**. Kingston-upon-Hull: Humber Archaeology
http://www.english-heritage.org.uk/publications/rczas-gibraltar-point-norfolk-phase2/

Norfolk

N1 Robertson, D, Crawley, P, Barker, A and Whitmore, S 2005 *Norfolk Rapid Coastal Zone Assessment Survey. Assessment and Updated Project Design*. Norfolk Archaeological Unit Report **1045**. Norwich: NAU
http://www.english-heritage.org.uk/publications/norfolk-rczas/

N2 Albone, J, Massey, and S. Tremlett, S 2007 *The Archaeology of Norfolk's Coastal Zone. Results of the National Mapping Programme*. English Heritage Project 2913. Gressenhall: Norfolk Landscape Archaeology/English Heritage

Suffolk

S1 Everett, L, Allan, D, and McLannahan, C 2003 *Rapid Field Survey of the Suffolk Coast and Intertidal Zone, May 2002-March 2003*. Ipswich: Suffolk County Council Archaeological Service
http://www.english-heritage.org.uk/publications/suffolk-rczas-assessment-report/

S2 Everett, L 2007 *Targeted Inter-tidal Survey*. Suffolk County Council Archaeological Service Report **2007/192**. Ipswich: Suffolk County Council Archaeological Service
http://www.english-heritage.org.uk/publications/suffolk-rczas-targeted-inter-tidal-survey-report/

S3 Good, C and Plouviez, J, 2007 *The Archaeology of the Suffolk Coast*. Ipswich: Suffolk County Council Archaeological Service
http://www.english-heritage.org.uk/publications/suffolk-rczas-archaeological-service-report/

S4 Hegarty, C and Newsome, S 2005 *The Archaeology of the Suffolk Coast and Inter-tidal Zone. A report for the National Mapping Programme*. Bury St Edmunds/Swindon: Suffolk County Council Archaeological Service/English Heritage
http://www.english-heritage.org.uk/publications/suffolk-rczas-national-mapping-programme-report/

Essex

E1 Heppell, E M and Brown, N 2001 *Greater Thames Estuary Essex Zone Monitoring Survey. Assessment and Updated Project Design*. Chelmsford/Braintree: Essex County Council
http://www.english-heritage.org.uk/publications/
essex-rczas-assessment-and-upd-2001-no-figures/

E2 Heppell, E M and Brown, N 2002 *Greater Thames Estuary Essex Zone Monitoring Survey. Interim report No. 1*. Chelmsford/Braintree: Essex County Council
http://www.english-heritage.org.uk/publications/
essex-monitoring-survey-interim-report-1-2002/

E3 Heppell, E M 2003 *Greater Thames Estuary Essex Zone Monitoring Survey. Interim report No. 2*. Chelmsford/ Braintree: Essex County Council
http://www.english-heritage.org.uk/publications/
essex-monitoring-survey-interim-report-2-2003/

E4 Heppell, E M, Brown, N and Murphy, P 2004 *Greater Thames Estuary Essex Zone Monitoring Survey. Assessment and Updated Project Design*. Chelmsford/Braintree: Essex County Council
http://www.english-heritage.org.uk/publications/
essex-monitoring-survey-final-assessment-and-upd-2004/

E5 Heppell, E M and Brown, N Undated *Rapid Coastal Zone Survey and beyond: research and management of the Essex coast*. Chelmsford/Braintree: Essex County Council
http://www.english-heritage.org.uk/publications/
essex-jwa-synthetic-paper/

North Kent

NK1 Wessex Archaeology 2000 *Historic Environment of the North Kent Coast. Rapid Coastal Zone Assessment Survey. Survey Phase 1. Final Report*. Ref: **46561**. Salisbury: Wessex Archaeology
http://www.english-heritage.org.uk/publications/
north-kent-coast-phase-i-1999-2000/

NK2 Wessex Archaeology 2002 *North Kent Coast. Rapid Coastal Zone Assessment Survey Phase II: Preliminary Field Investigation* Ref: **46564.01**. Salisbury: Wessex Archaeology
http://www.english-heritage.org.uk/publications/
north-kent-coast-phase-ii-pfi-2001/

NK3 Wessex Archaeology 2004 *North Kent Coast Rapid Coastal Zone Assessment Survey Phase II: Field Assessment: Pilot*. Ref: **46565**. Salisbury: Wessex Archaeology
http://www.english-heritage.org.uk/publications/
north-kent-coast-phase-ii-pilot-2002/

NK4 Wessex Archaeology 2004a *North Kent Coast Rapid Coastal Zone Assessment Survey. Phase II: Field Assessment 2003 Pilot Fieldwork*. Ref: **55057.02**. Salisbury: Wessex Archaeology
http://www.english-heritage.org.uk/publications/
north-kent-coast-phase-ii-pilot-2003/

NK5 Wessex Archaeology 2005 *North Kent Coast Rapid Coastal Zone Assessment Survey Phase II: Field Assessment. Year One Report*. Ref: **56750.02**. Salisbury: Wessex Archaeology
http://www.english-heritage.org.uk/publications/
north-kent-coast-phase-ii-year-1-2004/

NK6 Wessex Archaeology 2006 *North Kent Coast Rapid Coastal Zone Assessment Survey Phase II: Field Assessment. Year Two Report*. Ref: **56751.01**. Salisbury: Wessex Archaeology
http://www.english-heritage.org.uk/publications/
north-kent-coast-phase-ii-year-2-2005/

NK7 Wessex Archaeology 2005a *NKC Joint Fieldwork Report. June-July 2004*. Ref **56320.03**. Salisbury: Wessex Archaeology.
http://www.english-heritage.org.uk/publications/
north-kent-coast-planarch-2004/

NK8 Wessex Archaeology 2005b *NKC Planarch Participation. Essex Joint Fieldwork Report. May 2005*. Ref: **56321.02**. Salisbury: Wessex Archaeology
http://www.english-heritage.org.uk/publications/
north-kent-coast-planarch-2005/

South East

SE1 Wessex Archaeology 2011 *South East Rapid Coastal Zone Assessment Survey (SE RCZAS). Phase 1: National Mapping Programme Report – Blocks B, C, L and M*. Report Ref. **71330.01**. Salisbury: Wessex Archaeology
http://www.english-heritage.org.uk/publications/
serczas-phase1-nmp/

SE2 Cornwall County Council and Gloucestershire County Council 2012 *South East Rapid Coastal Zone Assessment Survey. National Mapping Programme. Components 1 and 2. NHPCP Project Numbers 6105 and 6106. Results of AP Mapping*. Truro/Gloucester Cornwall County Council and Gloucestershire County Council
http://www.english-heritage.org.uk/publications/
serczas-nmp-comp-1-2/

SE3 Wessex Archaeology. South East Rapid Coastal Zone Assessment Survey (SE RCZAS). Phase 1 Desk-Based Assessment. Ref. **71330.02**. Salisbury: Wessex Archaeology
The digital report will be online in the course of 2014.

Isle of Wight

IoW Isle of Wight County Archaeology and Historic Environment Service 2000 *Isle of Wight Coastal Audit*. Ventnor: Isle of Wight County Archaeology and Historic Environment Service
http://www.english-heritage.org.uk/publications/
isle-of-wight-coastal-audit/

New Forest

NF1 Wessex Archaeology 2010 *New Forest Rapid Coastal Zone Assessment. Phase I Desk-based Assessment. Main Report*. Ref: **72200.02**. Salisbury: Wessex Archaeology
http://www.english-heritage.org.uk/publications/
new-forest-rcza-phase-1/

NF2 Wessex Archaeology 2011 *New Forest Rapid Coastal Zone Assessment Survey. Stage II: Field Assessment*. Report Ref: **72201.1**. Salisbury: Wessex Archaeology.
http://www.english-heritage.org.uk/publications/
new-forest-rczas-phase-2/

NF3 Trevarthen, E 2010 *Hampshire Aggregate Resource Assessment: Aerial Photography Enhancement Results of NMP Mapping. English Heritage Project No. 5783*. Report No: **2010 R026**. Truro: Historic Environment Projects, Environment, Planning and Economy, Cornwall Council.
http://www.english-heritage.org.uk/publications/
hampshire-aggregate-resource-assessment-nmp/

Dorset

D1 Wessex Archaeology 2004a Historic Environment of the Dorset Coast. Rapid Coastal Zone Assessment Survey Phase I Project Report. Ref: **51958.05**. Salisbury: Wessex Archaeology
http://www.english-heritage.org.uk/publications/
dorset-rczas-report/

D2 Wessex Archaeology 2004b *Historic Environment of the Dorset Coast. Rapid Coastal Zone Assessment Survey Phase I Project Report*. Ref: **51958.06**. Salisbury: Wessex Archaeology
http://www.english-heritage.org.uk/publications/
dorset-coast-research-framework/

Isles of Scilly

IoS Johns, C, Larn, R and Tapper, B P 2004 *Rapid Coastal Zone Assessment for The Isles of Scilly*. Report No: **2004R030**. Truro: Historic Environment Service, Cornwall County Council.
http://www.english-heritage.org.uk/publications/
isles-of-scilly-rczas/

South-West

In the English Heritage *National Heritage Protection Plan* (Activity 3A2), work on Phase 1 of the SW RCZAS (South Coast: Devon, Dorset, part of Hampshire) is scheduled to begin in financial year 2012/13. The SW RCZAS North Coast (Land's End to Gore Point, Porlock Bay, Somerset), and Phase 2 surveys for both areas will follow as funding permits. Aerial photographic survey of Cornwall has been completed and that for the North Devon Area of Outstanding Natural Beauty (AONB) is underway.

Severn Estuary

S1 Mullin, D, Brunning, R and Chadwick, A 2009. *Severn Estuary Rapid Coastal Zone Assessment Survey. Phase 1 Report for English Heritage. (HEEP Project No. 3885)*. Gloucester: Gloucestershire County Council/Somerset County Council
http://www.english-heritage.org.uk/publications/severn-estuary-rczas-phase1/

S2 Catchpole, T and Chadwick, A M 2010 *Severn Estuary Rapid Coastal Zone Assessment Survey. Updated Project Design for Phase 2 Main Fieldwork for English Heritage (HEEP Project No. 3885)*. Gloucester: Gloucestershire County Council/Somerset County Council
http://www.english-heritage.org.uk/publications/severn-estuary-rczas-phase2a/

S3 Dickson, A, Catchpole, T and Barnett, L P 2010 Severn Estuary Rapid Coastal Zone Assessment Survey: Purton Hulks Aerial Photographic Progression Study (English Heritage Project No. 3885 2a PILOT). Gloucester: Gloucestershire County Council, Friends of Purton & English Heritage: Gloucester
http://www.english-heritage.org.uk/publications/severn-estuary-rczas-purton-hulks/

S4 Chadwick, A and Catchpole, T 2011 *Severn Estuary Rapid Coastal Zone Assessment Survey. Phase 2 Fieldwork Report (English Heritage Project NHPCP 3885)*. Gloucester: Gloucestershire County Council.
http://www.english-heritage.org.uk/publications/severn-estuary-rczas-phase2/

North-West

NW1 Johnson, B 2009 *North West Rapid Coastal Zone Assessment (NWRCZA)*. ARS Ltd Report **2009/53**. Gateshead: Archaeological Research Services Ltd
http://www.english-heritage.org.uk/publications/nwrcza/

NW2 Johnson, B 2009 *North West Rapid Coastal Zone Assessment (NWRCZA). Executive summary*. ARS Ltd Report **2009**. Gateshead: Archaeological Research Services Ltd
http://www.english-heritage.org.uk/publications/nwrcza-exec-summary/

NW3 Bacilieri, C, Knight, D and Williams, S 2009 *North West Coast Rapid Coastal Zone Assessment Survey. Air Survey Mapping Report. English Heritage National Mapping Programme. Historic Environment Enabling Programme: Project Number* **4548**. Gateshead: Archaeological Research Services Ltd
http://www.english-heritage.org.uk/publications/nw-rczas-air-survey-mapping-report/

NW4 Edie, G 2012 *The North West Rapid Coastal Zone Assessment (NWRCZA). Executive Summary Document*. Bakewell: Archaeological Research Services.

NW4 Edie, G 2012 *The North West Rapid Coastal Zone Assessment (NWRCZA)*. Bakewell: Archaeological Research Services.
http://www.english-heritage.co.uk/publications/nwrcza-phase2-project-report/

Introduction to the coastal historic environment

Where does the coast end? What is specifically characteristic about the coastal historic environment? Large areas of the country now inland were, at one time, coastal. Post-glacial marine to intertidal sediments extend inland in the East Anglian fens for almost 50km and nowhere in England today is more than 120km from the shoreline. Many places well away from the coast are easily reached along rivers that are tidal in their lower reaches. Besides this, coasts are dynamic but historic assets are fixed, and so sites that originally had nothing to do with the coast at all may end up there today as a result of later coastal change. This is especially true for the Palaeolithic (up to 1 million years to

11,500 years ago) and Mesolithic (c 11,500–6,000 years ago). Conversely, for much of those periods, other coastlines now lie well offshore (Fig 1.1).

It would be easy to write a text that includes almost *everything*, so we have to be more specific and rigorous. For practical purposes an arbitrary line has to be drawn somewhere. For the aerial photographic component of the Rapid Coastal Zone Assessment Surveys (RCZAS) all 1km National Grid Squares contiguous with the present coast – on land and on shore – were chosen, since this included the area where change will happen, and where decisions on management and conservation

Figure 1.1
Aldbrough, Holderness, East Yorkshire. Rotational failure of till cliff. This coastline has eroded significantly since the medieval period with the loss of 23 recorded settlements.

will have to be made (Fig 1.2). For other survey objectives a different study area is more appropriate, sometimes extending out to the 12 Nautical Mile limit.

Despite these arbitrary decisions there are three fundamental senses in which the coastal historic environment is distinctive. First, there are types of archaeological sites, buildings and landscapes that are confined to the coast and do not occur anywhere else. Examples include historic sea defences, fish and shellfish middens, marine fish traps, salterns, wrecks, ports, lighthouses, lifeboat stations, coastal military defences and coastal landscapes including grazing marsh. Secondly, there are archae-ological sites that are certainly present inland but are very rarely exposed and so usually difficult to study, for they are generally buried deeply beneath later deposits. Erosion on the coast makes them unusually visible and accessible, often providing extensive exposures; but they are also especially vulnerable, for their exposure makes them especially apt to destruction by natural processes. Examples include early prehistoric river channels, the sediments infilling meres, land surfaces, peats and ancient (now submerged) woodlands. Finally, there are formerly terrestrial sites now

under the sea, which may or may not have originally been on coasts, but have been submerged by rising relative sea level. Most are of Palaeolithic or Mesolithic date, but they also include sites as recent as parts of the medieval city of Dunwich, Suffolk.

These things will subsequently be referred to below as coastal historic *assets*, an ugly term, but one that is useful in terms of making the point that they are things of value, which we all own communally if not specifically.

Recording in the past

Observations of coastal and intertidal archae-ological and palaeoecological features date back at least to the 17th century, when Samuel Pepys recorded a buried prehistoric woodland during excavations for a dock at Blackwall in 1665: '… perfect trees over-covered with earth. Nut trees, with the branches and the very nuts upon them…'. The significance of such deposits was not established until the early 20th century by Clement Reid (1913), who appreciated their meaning in terms of wider coastal change: they represented land that had been sub-merged. Probably the first visual record of the

Figure 1.2
Benacre Broad, Suffolk.
A narrow shingle bank
separates the freshwater
Broad (on the left) from the
North Sea (to the right) and
it will eventually be over-
topped. Still-standing trees
have been killed by saline
groundwater.

archaeological effects of coastal erosion is the engraving by G J Newton, dated 1786, showing the collapsing walls of the late Roman fort at Walton Castle, Suffolk under the scrutiny of two gentlemen *cognoscenti* (Fig 1.3). Later port developments also produced archaeological material including remains of humans, red deer, aurochs, horse and whale, two prehistoric dug-out boats, a brushwood platform and bronze spear-head, all found during construction of the Preston Docks in the 19th century (Gonzalez and Cowell 2007). During construction of the Royal Edward Dock at Avonmouth in 1903 a Bronze Age rapier was recovered from a depth of some 50 feet (15.24m), apparently in channel sediments (Brett 1996). Very large quantities of prehistoric and later artefacts were recovered from the Thames during the 19th and early 20th centuries as a result of capital dredging for navigation (Lawrence 1929).

Other site-based coastal archaeological workers included Hume, working on the multi-period port site of Meols on the Wirral in the 1860s (Griffiths *et al* 2007), Spurrell (1889) who recorded intertidal peats and a Roman occupation surface in the Thames, and Reader (1911), who first recorded an intertidal

Mesolithic site in the River Crouch, Essex, although he believed it to be Neolithic. More extensive landscape-scale studies of the Mesolithic and Neolithic sites of the so-called 'Lyonesse surface' on the Essex coast were undertaken by Warren *et al* (1936). From the 1970s onwards new regional surveys were initiated, including the work of the Severn Estuary Levels Research Committee, published in successive volumes of its journal *Archaeology in the Severn Estuary*, a landscape study of prehistoric and later sites in the intertidal zone of Wootton-Quarr on the north coast of the Isle of Wight (Loader *et al* 1997; Tomalin *et al* 2012), a survey of intertidal sites in Langstone Harbour, Hampshire (Allen and Gardiner 2000), and a survey of the archaeology of the Essex coast (Wilkinson and Murphy 1995) (*see also* Fig 1.4).

The more modern surveys demonstrated the wealth of stratified intertidal archaeological material to be found and recorded; but they also highlighted that coverage of the coast in the then-National Monuments Record (now National Record of the Historic Environment, NRHE), and local authority Historic Environment Records (HERs) was inadequate in terms of providing an information base to

Walton Castle, Suffolk.

Figure 1.3
Walton Castle, Suffolk.
A later Roman fort eroding
in the 18th century. The
image was published in
1786, but relates to earlier
cliff erosion.

Figure 1.4
South Woodham Ferrers,
Essex. A section in the bank
of the Crouch Estuary,
showing the basal land
surface and overlying
sediments. (Ellen Heppell,
reproduced courtesy of and
© Essex County Council)

support responses from historic environment professionals to change induced by coastal management and development.

Alongside this, research into offshore submerged landscapes and coasts has expanded enormously. Research priorities were defined in Flemming (2004) and this was followed by the *North Sea Prehistory Research and Management Framework* (Peeters *et al* 2009). The latter emphasised the need for research on: stratigraphic and chronological frameworks; palaeogeography and environment; global perspectives on inter-continental hominin dispersals; Pleistocene hominid colonisations of northern Europe; re-occupation of Northern Europe after the Last Glacial Maximum; post-glacial land use dynamics in the context of a changing landscape; and representation of prehistoric hunter-gatherer communities and lifeways. Between 2002 and 2010 the Marine Aggregates Levy Sustainability Fund (MALSF) supported a series of projects that investigated the palaeogeography, Quaternary geology, palaeoecology and faunal remains of offshore submerged palaeolandscapes, as well as considering the contexts of related Palaeolithic and Mesolithic artefacts (www.marinealsf-navigator.org.uk/). As a consequence, our knowledge of submerged Palaeolithic and Mesolithic landscapes and sites has increased very significantly. A full review of the results is not appropriate here, but some results are discussed in Chapter 2.

Fulford *et al* (1997, 50–65) give a theoretical model of archaeological site locations and some specific examples in relation to changing coastlines. Rather than reiterating their discussion and examples, new results from recent studies at particular sites will be considered here. Types of sediment in which archaeological sites and finds occur – peats, estuarine silts, and coarser sediments such as sands and gravels – are reviewed for Southern England by Timpany (2009).

Beaches and dunes

Sand and shingle beaches are high-energy environments, subject to direct wave impact, where rapid destruction of exposed archaeological deposits and structures is likely. Unstratified artefacts are very commonly found on beaches, but they have mostly come from nearby sites under eroding dune systems, on cliff-tops or in horizontal foreshore exposures. They are useful principally as indicators of nearby eroding sites. An example includes the material recovered from Meols, on the tip of the Wirral, in the 19th century, which begins with three Carthaginian coins of the late 3rd century BC, followed by a chronologically wide range of artefacts dating through to the Middle Ages, indicating a significant trading site that remained in use for a very long time (Griffiths *et al* 2007). To give a more recently recorded example, the distribution of lower Palaeolithic artefacts and butchered animal bone found, often unstratified, on the beaches of Norfolk led directly to the investigation of the earliest known lower Palaeolithic site in the country at Happisburgh, dating back to around 850,000 years and perhaps before that. Concentrations of beach finds gave an indication of areas where stratified deposits might occur (Parfitt *et al* 2010 and pers comm).

Despite the high-energy conditions, some types of robust archaeological sites, for example the stone-built fish traps on the Severn shoreline, have survived long-term exposure on a beach, albeit becoming increasingly degraded (Chadwick and Catchpole 2010). Offshore submerged sites are often a source of wrecks and fragments of wrecks that drift onto the beach as they break up (Fig 1.5); yet more-or-less intact wrecks – for example that of the largely buried *Amsterdam* (1749) on Bulverhythe, Hastings, or the exposed 19th-century ice-carrier *Vicuna* on Holme Beach, Norfolk, still survive. Apart from those associated with sea-defence, few structures have been built directly on a beach, but the Low Lighthouse at Burnham-on-Sea, Somerset (1832) is one example. It is a timber structure, supported by piles: changes in beach conformation may mean that it is not sustainable, in its present location, indefinitely (Fig 1.6).

Coastal dunes occur widely around the English coast (Fulford *et al* 1997, fig 31), with concentrations in the North-East, parts of East Anglia, the South-West and Lancashire. Dune sand is derived from offshore sediment sources or cliffs, and finally from the beach itself, and phases of dune formation and depletion are related to availability of off-shore sediment, changes in beach levels, the frequency and intensity of storms, and to climate change in general. Younger dunes typically occur closest to the shore, with older ones behind, often illustrating changes in vegetational succession from yellow to grey dunes, maintained by species such as *Ammophila arenaria* (marram grass), other halophytes, mosses, lichens and, ultimately, maritime scrub and woodland. However, dunes are ultimately unstable owing principally to physical disturbance by humans or grazing animals, wind erosion and sea-level rise, which initiate erosion and dune migration. From an archaeological point of view their key significance is that they seal buried soils, landscapes and sites, protecting them from subsequent disturbance; but also that dune

movement can leave sites in a very different environmental context from that in which they were constructed. Bell and Brown (2008) provide a database of dune systems and associated archaeological sites and artefacts in southern England and illustrate the contexts in which material may occur. Fulford *et al* (1997, 65) note examples including Bronze Age settlement and later fields at Gwithian, Cornwall, and the 30m-thick dune sequence including at least five phases of Bronze Age occupation at Brean Down, Somerset. Since then, two sites in particular have received detailed examination.

Figure 1.5
Caister-on-Sea, Norfolk. Part of a large 18th- or 19th-century timber vessel on the beach, eroded from an offshore wreck.

Figure 1.6
Burnham-on-Sea, Somerset. The Low Lighthouse (1832).

The Bronze Age timber circuit surrounding a felled inverted oak excavated in 1998–9 at Holme-next-the-Sea, Norfolk, lay on a beach seawards of an existing dune system (Brennand and Taylor 2003). The site – commonly known as 'Seahenge' (or Holme I) – was constructed (on dendrochronological evidence) in the spring or early summer of 2049 BC (Fig 1.7). As exposed in 1998 the site was related to a sequence of peats and intertidal clays. Plant macrofossils and insect remains (*see* section on Environmental Archaeology in Chapter 2) from archaeological contexts included a high proportion of salt-marsh, upper mudflat and dune species. Pollen analysis registered inputs from woodland on higher ground inland, as well as more local salt-marsh and grassland, while ostracods and foraminifers also indicated mudflat environments. Taking together the evidence from these environmental indicators, with the sequence of sediments and dating evidence at the site, Murphy and Green (2003) proposed a model for local landscape change. Before the monument was constructed the local area comprised mudflats and salt-marsh, behind a low, developing dune system, which was often over-topped by the sea. As the dunes increased in height and stabilised, tidal influence was reduced and the extent of salt-

marsh increased, providing a semi-terrestrial walkable environment in which the monument was built. Further dune development later isolated the area from the North Sea, permitting development of reedswamp and alder wood-land, beneath which a peat developed. From around 1300 cal BC, and into the present day, the dunes migrated inland over, and ultimately beyond, the monument, leaving it isolated on the modern beach. It was, therefore, *not* constructed on the beach – contrary to much popular opinion – but in an area of salt-marsh lying between a dune system and woodland on higher ground. It would have been con-spicuously visible in such an open habitat, and was part of a wider complex of monuments. Given the site's present location on a high-energy beach, its ultimate destruction was inevitable and so a programme of excavation (of Holme I) and subsequent monitoring and recording of other parts of the site as they eroded was undertaken (Ames and Robertson 2009).

The site at Low Hauxley, Druridge Bay, Northumberland, currently underlies a dune system designated as a Site of Special Scientific Interest (SSSI) (NE3, NE6). It is on a buried natural hillock in the pre-dune land surface, with organic sediment units to the north and south, including peats, dating broadly to the 4th millennium BC onwards (Fig 1.8). The archaeological site comprises an eroding Mesolithic soil surface (including lithics and faunal remains), grading downhill into peats, with a Beaker to early Bronze Age cemetery at the same location, including inhumations and cremations in cists. However, erosion of the dune front, combined with extensive open-cast coal mining in the land behind the dunes has meant that the surviving area of the site is restricted and eroding. Recent monitoring at Low Hauxley has shown erosion rates of 0.5–1.5m annually and this rate is likely to increase as relative sea-level rises. The Shoreline Management Plan option for the dunes is 'Managed Realignment' with some localised active management of drainage and access (*see* Chapter 6). Continuing erosion since the 1980s has resulted in exposure and then destruction of archaeological deposits, and there has been intermittent excavation and survey over some 30 years, most recently comprising recording of Bronze Age cremations in 2009, and of a peat on the beach with human and animal footprints and worked timber on its

Figure 1.7
Holme-next-the-Sea,
Norfolk. A Bronze Age
timber structure, known
as 'Seahenge'.
(Reproduced by permission
of NPS Archaeology)

Figure 1.8
Low Hauxley, Druridge Bay,
Northumberland.

surface in late 2010. Overall, the site was assessed in the RCZAS report as being of national significance, both for its Mesolithic and Bronze Age components and for the associated palaeoenvironmental evidence. Monitoring the site, with intermittent excavation, has produced a fragmented record (largely unpublished) with poor spatial information. English Heritage has provided funds for a thorough review and appraisal of earlier work and for the recording of the exposed surface that shows footprints. However, controlled area excavation of part of the site is plainly required, and would fit with Northumberland Wildlife Trust's intended management of the SSSI to breach the dunes and allow salt-water lagoon formation behind them.

Estuaries, salt-marshes and mudflats

Intertidal environments in sheltered embayments, estuaries, intertidal mudflats and salt-marshes provide different conditions, generally speaking being lower energy environments, though with erosion at some locations (Environment Agency 2011). Salt-marsh supports highly significant natural communities of plants and animals and also provides 'ecosystem services', primarily in terms of reducing wave energy, so that tidal defences behind the marsh have a degree of protection. For present purposes the key fact is that salt-marsh sediments include, or overlie, archaeological deposits.

Among the more significant deposits in these environments are peats, typically found interstratified with clayey or silty (minerogenic) sediments. Locations of peats, submerged forests and other coastal sediments known in 1997 are shown in Fulford *et al* (1997, fig 26). Since then there have been numerous further records, made during the RCZAS and other projects. Data on locations, general descriptions, altitudes relative to modern sea level, associated archaeological material, radiocarbon dates and bibliographic references have been incorporated into a country-wide database (Hazell 2008 and http://www.english-heritage.org.uk/professional/research/heritage-science/environmental-studies/Environmental-Studies-Resources/intertidal-peat-database/). The database contains upwards of 300 entries and is continually updated. Sites range in exposure from peats visible on many intertidal surfaces through to fully submerged Mesolithic peats, as at Bouldnor Cliff off the north-west of the Isle of Wight (Momber *et al* 2011); and in date from the Ipswichian to Holocene. Local changes in coastal morphology have, in many places, permitted peat formation behind natural barriers at elevations at or below contemporary sea level. Dendrochronology (*see* Chapter 2) has permitted precise dating of some submerged forests, for example at Quarr Beach (3463 to 2557 BC) (Fig 1.9), and Fishbourne on the Isle of Wight (3280 to 1980–1750 BC (Tomalin *et al* 2012, 60–79).

Other sites and landscapes pre-dating marine transgression in estuaries and embayments include locations where submerged land surfaces are sealed beneath marine and intertidal sediments. They originated as dryland sites, of Neolithic or earlier date, and consequently preservation of organic materials is limited, comprising primarily pollen, charred plant material and bone. The submerged Neolithic landscape of The Stumble, Blackwater Estuary, Essex (discussed further in Chapter 4), is typical. During the early to middle Neolithic, from 3685–3385 cal BC, the area was low-lying land, around 1km from the nearest tidal creek, drained by freshwater streams (Fig 1.10). Local primary woodland was dominated by lime, oak and hazel. There were small-scale woodland clearances associated with farming and exploitation of wild foods and other resources. As relative sea level rose, the Blackwater Estuary expanded and, by the later Neolithic, soils in the vicinity were becoming waterlogged, freshwater streams were becoming tidal creeks, and a zone of salt-marsh expanded progressively inland (Fig 1.11). Rising groundwater resulted in the death of trees at the site, and ultimately preservation of their root systems. The latest evidence of human activity on the submerged land surface came from a 'burnt flint mound', dated to 2490–2285 cal BC. Later, during the Bronze Age, rising Relative Sea Level resulted in the area becoming salt-marsh and mudflat. Wooden structures, probably related to salt-marsh grazing and fishing, were constructed, and settlement sites were relocated to the dry soils of the adjacent gravel terraces. From the Iron Age onwards salt-marsh and mudflat resources at the site continued to be exploited (Fig 1.12). Continued RSL rise and erosion resulted in extensive foreshore exposures of sites of all periods (Wilkinson *et al* 2012) (Fig 1.13).

Figure 1.9
The shore between Wootton and Quarr, Isle of Wight, showing a submerged forest.

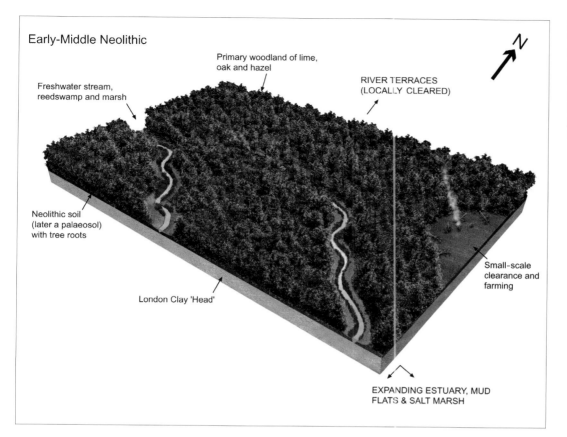

Early-Middle Neolithic

Primary woodland of lime, oak and hazel

RIVER TERRACES (LOCALLY CLEARED)

Freshwater stream, reedswamp and marsh

Neolithic soil (later a palaeosol) with tree roots

London Clay 'Head'

Small-scale clearance and farming

EXPANDING ESTUARY, MUD FLATS & SALT MARSH

N

Figure 1.10
The Stumble, Blackwater Estuary, Essex. Palaeogeography in the earlier Neolithic. (Images redrawn by Ian Bell from drafts prepared by the author, © Essex County Council)

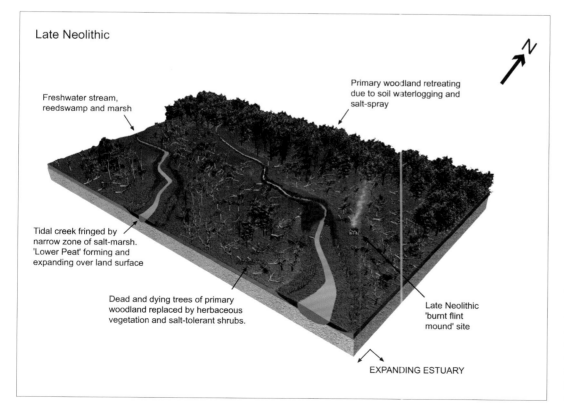

Late Neolithic

Primary woodland retreating due to soil waterlogging and salt-spray

Freshwater stream, reedswamp and marsh

Tidal creek fringed by narrow zone of salt-marsh. 'Lower Peat' forming and expanding over land surface

Dead and dying trees of primary woodland replaced by herbaceous vegetation and salt-tolerant shrubs.

Late Neolithic 'burnt flint mound' site

EXPANDING ESTUARY

N

Figure 1.11
The Stumble, Blackwater Estuary, Essex. Palaeogeography in the later Neolithic. (Images redrawn by Ian Bell from drafts prepared by the author, © Essex County Council)

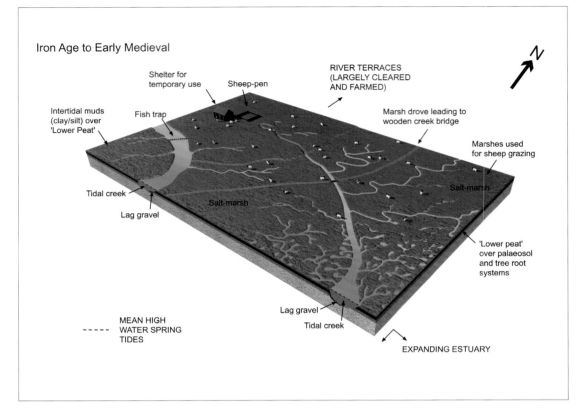

Figure 1.12
The Stumble,
Blackwater Estuary,
Essex. Palaeogeography
in the Iron Age to
medieval periods.
(Images redrawn by
Ian Bell from drafts
prepared by the author,
© Essex County
Council)

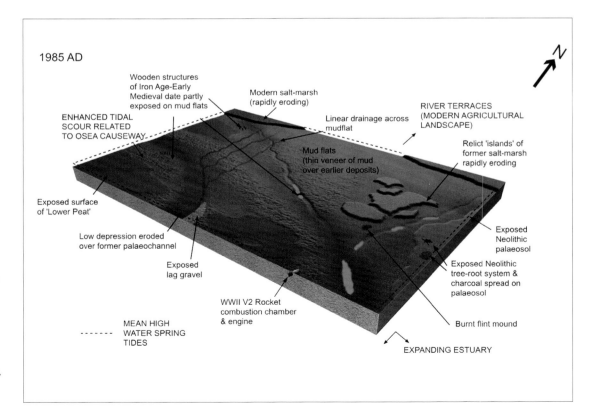

Figure 1.13
The Stumble,
Blackwater Estuary,
Essex. The site in 1985,
when first discovered.
A German V2 rocket
landed on the site
in 1945.
(Images redrawn by
Ian Bell from drafts
prepared by the author,
© Essex County
Council)

Figure 1.14
The Nass, Essex. A typical
V-shaped timber fish trap.
Radiocarbon dating shows
that the majority of similar
structures in this area are
of middle Saxon date.
(Photograph taken by
the late Ron Hall)

Once intertidal habitats were widely established land-use in the estuaries shifted towards more specifically maritime activities. Stationary fishing structures, including timber and stone fish traps (mainly of the early medieval period) and salterns (from the Bronze Age onwards) were developed in many areas (Figs 1.14–1.15). Hulks, related to coastal trade, survive well, interstratified within estuarine deposits (Davies 2011). Construction of sea-walls, counter-dykes, drainage ditches and sluices, is thought to have originated in the Severn Estuary in the Roman period, though more generally date to the early medieval period (Fig 1.16). Sites of these types are considered in more detail in Chapter 4.

Figure 1.15
Alde Estuary, Suffolk. A late
Iron Age to early Roman Red
Hill (saltern) seen in section.
(Reproduced by permission
of Suffolk County Council
Archaeological Service)

Figure 1.16
Clifton Marshes, Lancashire.
A relict sea bank left in
grazing marsh after new
sea defences were built
further seawards.

Cliff coastlines

Cliffs continually erode, though the rate of loss is determined primarily by their geology, exposure to the prevailing wave climate, levels of beaches and freshwater flow through them. The archaeological solution to cliff erosion is, plainly, excavation and recording before loss but funding to achieve this may be problematic (*see* Chapter 6). In Suffolk, the major medieval town of Dunwich, situated on soft cliffs of glacial sand, has been almost entirely lost (http://www.dunwich.org.uk/). Erosion is recorded as early as the Domesday Book with further losses reported during the storms of 1328 and 1347. Erosion continued throughout the 16th century, and storms in 1740 destroyed large areas of the rest of the city, so that only All Saints church remained, finally falling from the cliff in 1919 (Fig 1.17). Current geophysical and diving work offshore has found remains of large stone structures, including churches, surviving at approximately their original locations (Sear *et al* 2011). Although structures on the higher part of the town have largely been destroyed, lower lying parts, including the harbour, may well survive intact. New work will focus on this area. It will also collect higher resolution imaging of the existing ruins. Additionally, it is hoped that data from

Figure 1.17
Dunwich, Suffolk. The
surviving tower buttress
of All Saints, taken down
before it fell over the cliff
edge and re-erected in
St James' churchyard
in 1922.

historical bathymetric charts and pilot books will increase understanding of changes in the seabed and the exposure or burial of structures.

The decision to give the South-West a low priority in the RCZAS, which was largely due to the extensive areas of hard-rock resistant cliff, which erode slowly, masks one very significant problem: the increasing vulnerability of places at the toes of cliffs near sea level. Near-shore dredging for aggregates resulted directly in the loss of the village of Hallsands, Devon, in the early 20th century by a process of beach 'draw-down' Fulford *et al* (1997, 147) (Fig 1.18). Aggregate dredging will never have the same effect again owing to modern regulation, which confines dredging to offshore areas with no impacts on the coast. However, there are many harbours in the South-West, at the foot of cliffs and in low-lying inlets, in risky places. Many were established to serve specific economic functions, frequently the export of mineral ores or China Clay, or explicitly for fishing. While they were economically active, and part of a profitable business, continual maintenance to repair damage resulting from their exposure to the Atlantic storms was part of the deal. Later, deprived of their primary purpose, most reverted to being minor fishing ports and tourist destinations (Fig 1.19). Neither activity now

Figure 1.18
Hallsands, Devon. Remains of the village, destroyed by beach draw-down in 1917.

Figure 1.19
Charlestown Harbour, Cornwall.

produces funds that can support many of them. For example, Mullion Harbour on the Lizard is owned and managed by the National Trust. The harbour wall of 1895 was not constructed in an ideal location; even launching the lifeboat was often impossible due to storms. The Trust's intention now is to preserve the harbour for as long as is practically possible and then to realign the coast. This will involve abandoning further maintenance and then removing the breakwaters, or allowing them to collapse, while attempting to retain the inner harbour walls. In the longer term the site will revert to being unconstrained coastline. The next seriously damaging storm will determine the time-table. Elsewhere economically active ports, in general, are undergoing rapid change at present (Fig 1.20). In some cases this results in expansion, for example at Dover, Felixstowe, Harwich and Avonmouth; elsewhere ports are contracting and being redeveloped. Either way, decisions on the future of extant historic structures have to be made (see Chapter 6).

In many places World War II cliff-top military defences are among the first sites to be eroded. For example, the site of the World War II coastal searchlight or gun emplacement at Lee-on-the-Solent, Hampshire, was photographed from the air between 1942 and 1986; but by the latter date no trace of the site was visible, owing to

cliff erosion (SE1, 58). In the south-east of Kent, which was most exposed to enemy attack, and included what was then called 'Hell-Fire Corner' around Dover, over 60% of aerial photographic records created or enhanced for the NRHE dated to World War II (SE1, 50). Similar results were obtained for many other areas of the country, especially in the south and east. They are discussed further below, but range in scale from the massive Godwin Battery at Kilnsea on the Humber, which has now been almost totally destroyed by erosion, with most significant structures now lying collapsed on the beach, through to numerous collapsed pillboxes and other smaller defences. Elsewhere some defensive structures of this period are still extant but displaced, surviving intact as recalcitrant concrete structures – but not where they were constructed. A pillbox at Happisburgh, Norfolk, originally on the cliff top, now lies inverted on the beach, a considerable distance from the modern cliff line (Fig 1.21).

Coastal and offshore landscapes

Coastal grazing marshes, reclaimed from salt-marsh and mudflat, are a distinctive, but generally unappreciated coastal landscape (Fig 1.22). They include numerous archaeological

Figure 1.20
Sheerness, Kent. Georgian buildings in the Garrison Fort. They are still in use, and will continue to be so, as the Naval port is regenerated.

features such as saltern sites, sea walls, stock management features, and WW II military defences, as well as historic built structures, such as farmhouses and barns. Marshland landscapes in their present form developed from the Roman period onwards, involving phases, not necessarily sequential, of the exploitation, modification and transformation (conversion to agricultural production) by the construction of sea defences and drainage systems (Rippon 2000). A few marshland areas were reclaimed in the Roman period, notably in the Severn estuary, but coastal wetlands were more extensively recolonised from the 8th century AD onwards and on through the Middle Ages and early modern period: indeed in a few areas up to the 1970s. However, during a stormy climatic phase from the 1280s through to the mid-15th century breaches in flood banks were frequent, leading to losses of farmland. These were often not brought back into production for very long periods due to economic constraints (Galloway 2009). At Fambridge, Essex, there are remnants of breached sea walls destroyed in a storm surge of 1897, resulting in a buried soil and Victorian rubbish dumps being covered by intertidal sediments (Wilkinson and Murphy 1995, 208–9). This particular loss was never reclaimed. The extent of coastal agricultural land has therefore always been fluid. Land-claim was often not a one-way process.

In 2008, Palaeolithic material, including hand-axes, flakes, cores and faunal remains were recovered from dredging licence Area 240 (licensed to Hanson Aggregates Marine Ltd) about 11km off the coast of Great Yarmouth, Norfolk. The finds showed that stratified archaeological material can survive in deposits, originally of terrestrial origin, being extracted

Figure 1.21
Happisburgh, Norfolk. This WW II pillbox was originally on top of the cliff.

Figure 1.22
Halvergate Marsh, Norfolk. Grazing marshes are among the most threatened coastal habitats around the country.

Figure 1.23
Palaeolithic artefacts from
aggregate extraction Area
240, off the East Anglian
coast, collected by Jan
Meulmeester at an aggregate
wharf and displayed at
a meeting in Amersfoort
in 2008.

for aggregates (Fig 1.23). This led to a programme of offshore geophysical and geo-technical work, palaeoenvironmental studies, grab sampling and, finally, inspection of dredgings from the area. The results are being prepared for formal publication, but it is clear that, although artefacts of varying date and origin are present, most can be attributed to sediments deposited during the Wolstonian (Marine Isotope (MIS) Stage 8/7). These sediments infill a channel, probably cut at the end of the Anglian Glaciation (MIS 12), and form the floodplain of the channel (Wessex

Archaeology 2010; 2011). In effect, this work was the first attempt to develop a set of field techniques providing the stratigraphic control of an offshore excavation.

Large-scale land-claim has resulted in former coastal landscapes now lying very well inland, in the Fens of eastern England, Romney Marsh and other areas. However, to include such locations in the present publication would stretch the definition of the study area in the first paragraph of this chapter just too far. The focus will primarily be on the historic environment of the present coast and on its future.

Survey, recording and characterisation in the coastal zone

Fulford *et al* (1997, 74–102) give an outline of coastal survey methodologies, much of which still applies. The intention here is not to provide complete technical details of methods used in all the individual RCZAS surveys, for they are given in full in the reports (http://www.english-heritage.org.uk/professional/advice/advice-by-topic/marine-planning/shoreline-management-plans/rczas-reports/), but to outline in broader terms how methodology has developed since 1997.

Contributors to one of the immediate precursors of the Rapid Coastal Zone Assessment Survey programme – the Hullbridge Survey in Essex – literally walked, in the 1980s, around most of the coast of Essex – recording sites, and undertaking small-scale excavation (Wilkinson and Murphy 1995). The survey was focused on archaeological remains, mainly prehistoric to medieval features, apart from brief consideration of post-medieval embankments and breaching. Some types of later historic assets, such as hulks or 20th-century military structures, were not considered *at all*. The survey was not supported in advance by a desk-based study of aerial photographic sources and LiDAR (Light Detection and Ranging) and satellite imagery data were not then available. Techniques for recording site location were optically based, relying on a prismatic compass sighted on church towers and other conspicuous features, for Global Positioning Systems (GPS) devices had not then been developed for general use. Offshore surveying methods, using geophysics, were also not considered, for these, too, remained new, high-tech, and certainly not available to more than a tiny minority of maritime archaeologists. Nevertheless, environmental archaeological techniques were extensively applied, and a full programme of radiocarbon dating was deployed. The survey programme was followed immediately by a large-scale excavation project, at the Neolithic site of The Stumble in the Blackwater Estuary (Wilkinson *et al* 2012). Advances in technology, but perhaps more importantly in terms of the perception of what the historic environment comprises, and what should be done about threatened sites, have meant that more recent survey and subsequent studies have been very different. There is no substitute for walking, seeing and recording during field survey, as Flett (1937) emphasises in the case of the BGS, but much else has changed.

The Rapid Coastal Zone Assessment Surveys

The first Rapid Coastal Zone Assessment Surveys (RCZAS) were county-based, beginning in Suffolk. Subsequently, larger regional surveys were found to be more cost-effective, and better related to the Environment Agency's and Defra's Flood and Coastal Erosion Management programme (FCERM). These later surveys involved successively the Severn Estuary, Yorkshire and Lincolnshire, the North-East, North-West and South-East of England (Fig 2.1). They comprised two phases:

1. Phase 1 (desk-based assessment). This drew on data from aerial photographs, LiDAR, historic maps and charts, the local authority Historic Environment Records (HERs), and the NRHE. Data were captured in a Geographic Information System (GIS) with supporting databases to nationally agreed data standards.

2. Phase 2 (field assessment). This comprised a rapid walk-over survey, designed to verify records from Phase 1, locate and characterise site types not visible from the air, and assess significance and vulnerability.

Figure 2.1
The Rapid Coastal Zone
Assessment Surveys.
(Image redrawn by John
Vallender from a draft
provided by the author)

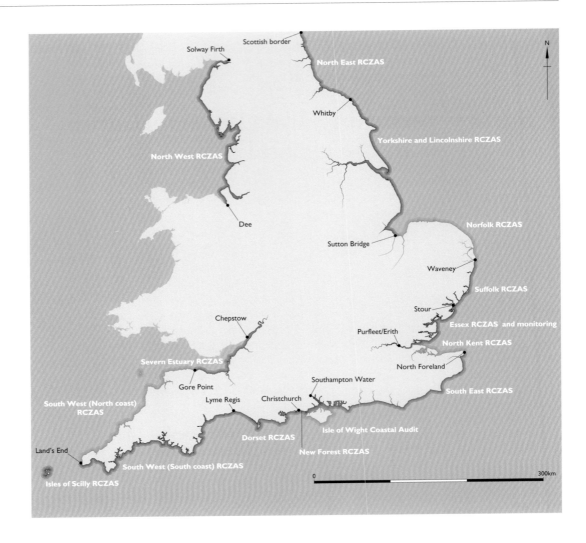

A Phase 3 was never formally defined, for it was regarded as a potentially expensive diversion from the primary aim of completing basic, outline national coverage. However additional work, especially scientific dating and more detailed survey, has been necessary to characterise sites fully in several parts of the country.

In 2011–12 the RCZAS were incorporated into EH's National Heritage Protection Plan (English Heritage 2012a) under the Activity 3A2 (coastal survey). Limitations of funding mean that full national survey, including the south-west of England will not be completed until after the current plan period, after 2015. The south-west coast was given a low priority in the programme, since so much of it comprises hard rock cliffs, where coastal change will be slow. However, some sections of this coast are much more dynamic. In some of these areas information is supplied, for example by the

Cornish Estuaries Audits (Ratcliffe 1997; Parkes 2000). Elsewhere, specific local studies, such as that at Poole Harbour (Dyer and Darvill 2010), provide useful data.

The outputs consist of enhanced HER and NRHE records, together with client reports for English Heritage. The information gained will enable us to make a better-informed input to the FCERM process, and will help to ensure effective mitigation of the effects of coastal change through the 21st century. It will also provide a database for use in further research and in the development control process. The results are considered below.

The surveys were guided by a series of Briefs, developed from 1999 onwards, and the work was commissioned from archaeological consultants. Each survey raised new issues that had previously not been considered and so the Briefs became progressively more comprehensive and more specific.

Desk-based survey (Phase 1)

For the Phase 1 desk-based assessment, the following main sources were consulted:

- Sites and Monuments Records (SMRs) or HERs (including National Trust and National Parks HER/SMR, where appropriate);

- The NRHE;

- Lists of Scheduled Ancient Monuments, Designated Wrecks, Listed Buildings, Conservation Areas, Registered Parks and Gardens, and Historic Battlefields;

- *Futurecoast* (Defra 2002) and the relevant Shoreline Management Plans (SMPs), and their supporting documentation. Particular attention was paid to the management option identified for each management unit and the projected rate of coastal change (*see* Chapter 6);

- Any other available studies of palaeogeography, coastal change, or historic map regression studies;

- Aerial photographs held by the NRHE, the University of Cambridge, the Environment Agency, Coastal Observatories and local authorities;

- Historic map and charts, held by the UK Hydrographic Office (UKHO), the National and County Record Offices, and other regional collections, and digitised early Ordnance Survey (OS) editions;

- Modern topography and bathymetry, derived from OS and UKHO data;

- The Portable Antiquities Scheme database; and

- Contacts with local individuals, societies and organisations.

Additional sources were often also consulted, including:

- Client reports for developers not available in the SMR/HER;

- Databases developed for thematic projects, such as The Defence of Britain and England's Shipping;

- Museum archives;

- Information on 'wreck' (recovered artefacts) declared to the Receiver of Wreck;

- UKHO Wreck Data; and

- Local Authority Maritime Archaeological Databases, where these existed.

However, by far the largest component of Phase 1, in terms of resources and cost, was aerial photographic transcription and digitisation. It was plain from the outset that extensive work intended to characterise landscapes more fully would not be possible in this project. Instead of the usual pattern of examining entire OS 1:10,000 Quarter Sheets (25 sq km), all 1km National Grid Squares contiguous with the present coast – on land and on shore – were defined as the study area. This was obviously limiting, but necessary, so as to focus attention on areas where change would happen and where management decisions would eventually be needed. Among the disputed factors during the development of project designs for each survey was how far inland – up estuaries and over low-lying reclaimed land – survey should extend. There was an inevitable tension between the objectives of the National Mapping Programme (NMP) – which was intended to provide an objective, considered, assessment of landscape blocks – and the RCZAS – which was more management-based. Practicality and the availability of funding dictated the decision. Standard NMP methodologies were followed, involving digital transformations of photographs to achieve rectification and the use of AutoCAD to trace off archaeological features, alongside the production of a monument record for each site mapped. The mapping provides a comprehensive picture of the resources present, though on some extensive intertidal flats, lacking any conspicuous easily located geographic features, precise location of sites and structures was a problem (Figs 2.2–2.3).

LiDAR is an aerial survey technique that can detect very small height variations on land surfaces, and can produce high-resolution digital terrain models. These can be given artificial 'shadows' by digital manipulation, which helps in the detection of ploughed-down earthworks that are invisible by other methods. As part of the Severn Estuary RCZAS the flood bank known as the Great Wall of Elmore, Gloucestershire, was studied using LiDAR data supplied by the Environment Agency (Fig 2.4). This extended its known length by some 30m and in the same area a previously unknown barrow or windmill mound, a flattened moated site, and ridge-and-furrow fields were mapped. Parts of fish-weirs off the Somerset coast were newly-recorded by LiDAR, while in the New Forest LiDAR was found to be helpful in defining

Figure 2.2
Blue Anchor Bay near
Minehead. A plot from an
aerial photograph of 1969,
showing offshore features
including fish traps
(Chadwick and
Catchpole 2010, fig 4).
(Reproduced by permission
of Gloucestershire County
Council Archaeology
Service)

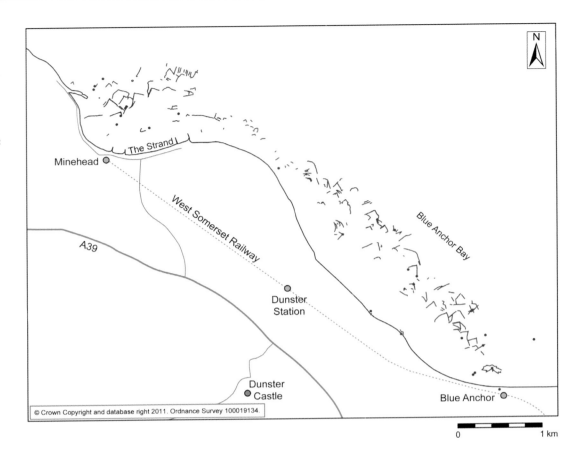

© Crown Copyright and database right 2011. Ordnance Survey 100019134.

Figure 2.3
Dunster, Somerset.
A stone-built fish trap
during field recording.

features such as sea-walls, salt works and extractive pits (Truscoe 2007; NF3, 16). On-line imagery can also be useful, including the images included in Google Earth (earth.google. com), on which hulks and other large structures are often plainly visible. The images available from the Channel Coast Observatory (www. channelcoast.org) are geo-referenced, which proved exceptionally useful for locating sites lying within large expanses of featureless mudflats, notably in Chichester Harbour, which had no other form of rectification. Moreover these recent images provided information on the current state of sites first recorded in earlier aerial survey (SE2, 3).

Aerial photographs are of no value for detecting prehistoric sites pre-dating, or unrelated to, earthwork or large-scale timber construction. Artefact scatters of all periods on foreshores are invisible from the air. Even sites such as the Bronze Age timber structures at Holme-next-the-Sea, Norfolk, which were previously known, could not be seen in aerial photographs. But elsewhere, large intertidal timber structures, such as Anglo-Saxon fish traps in the Essex estuaries, were plainly visible (Strachan 1995). In the Severn, differentiating lines of net hang clearances from other, more solid, components of fish traps was not always easy, and full interpretation depended on ground survey (Chadwick and Catchpole 2010).

Sites detected during the aerial photographic component of the RCZAS include a relatively high proportion of World War II military structures: on the coastline between Shoreham-by-Sea and Folkestone 92.5% of sites recorded were of this date (SE1, 50; SE2, 3). These sites are conspicuous, often involving concrete construction, though including earthworks and more temporary features, all plainly visible in 1940s aerial photographs. Some are no longer extant, owing to coastal erosion or later dismantling, notably at resorts such as Eastbourne, Bexhill, Folkestone and Brighton.

The overall strength of aerial photographic work is the long time frame, with photographs dating back to the 1940s, permitting recording of changes to sites through time. Detection of surviving remains of the 16th century Black Joy Forte, Norfolk, in photographs from the 1970s, permitted more detailed assessment of the site's survival on the ground (N2, 227–31). This advantage has been exploited for specific studies. Changes in the assemblage of some

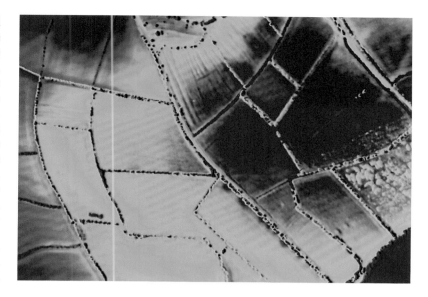

65 hulks at Purton (Fig 2.5), beached to provide coast protection, were assessed from a range of photographs dating from 1945–7 through to 2007. Successive images showed the addition of vessels and the degradation, or obscuring by sediment, of others. This information will aid future management of this hulk assemblage (S3).

Figure 2.4
The Great Wall of Elmore shown in Environment Agency LiDAR captured at 2m ground resolution (1 hit per 2m²). The image has been processed into a surface and lit from the west, additionally showing height range (c 5m–12m) from blue to red. (Image produced from Environment Agency tile SO7614 flown between Dec 2005 and Jan 2006 © English Heritage)

Figure 2.5
The hulks at Purton, Somerset, in 1969. (OS/69117 156 18-APR-1969 © Crown Copyright. Ordnance Survey)

Figure 2.6
The workboat used
during the North Kent
RCZAS on the Medway.
(© Wessex Archaeology)

Field survey (Phase 2): access and recording

Field survey, though guided by aerial photographic information, permits more detailed examination of smaller components of the historic landscape, which simply cannot be seen from the air. The purpose of the field assessment phase is to:

- Verify identifications made during the desk-based assessment;

- Locate and characterise sites and features undetected by the desk-based assessment;

- Determine the geomorphological and sedimentary context for features;

- Assess whether features are actively eroding;

- Selectively sample features; and to

- Test fieldwork methodologies and assess the practicalities and logistics of future fieldwork.

The first difficulty is often access. Where extensive areas above High Water Mark are to be examined, land access issues may be a serious constraint on survey, necessitating contact with landowners at an early stage, notably in terms of Military Ranges and Training Areas and sites with designations for the Natural Environment, such as SPAs (Special Protection Areas) and SSSIs (Sites of Special Scientific Interest).

Other constraints requiring consideration are intermittent exposure of intertidal sites; for example only at extreme low tides, after scour, or when not covered by algal growth. Walking is plainly cheapest and easiest, but may not always be possible or safe. The isolated mudflat and salt-marsh islands in the Medway could only be reached by boat, so an inflatable Avon workboat, with a shallow draft, an aluminium deck and large tube size, was used (NK 4) (Fig 2.6). Some extensive tidal flats in the Severn could not be crossed safely on foot (S4, 28-9). To access sites the survey team initially travelled on the Burnham-on-Sea Area Rescue Boat hovercraft during training exercises, but more extensive use of a hovercraft proved to be prohibitively expensive. Instead, an Argocat 8×8 tracked semi-amphibious all-terrain vehicle was hired (Fig 2.7). Although this

Figure 2.7
Argocat on mudflats in
the Severn Estuary.
(Reproduced by permission
of Gloucestershire County
Council Archaeology
Service)

proved to be unreliable on soft muds, used judiciously and on firmer surfaces, it sub-stantially speeded survey and reduced travelling time to and from sites.

Locating the positions of sites within the featureless terrain of tidal flats is now based on Global Positioning Systems (GPS) rather than optical equipment. GPS depends upon signals from four or more GPS satellites to determine location and elevation. Loss of signals may cause problems, at least intermittently, and especially near cliffs. For each site identified during the survey, an accurate co-ordinate is taken using a GPS with a differential correction in order to improve the accuracy of data to ±3m: though, in fact modern devices permit position-fixing horizontally to sub-metre accuracy (Fig 2.8). However, determining Z co-ordinates (elevation) has proved more problematic. Global positioning systems are a cost-effective method of gathering co-ordinates quickly and accurately, and also remove the need for mapping points by measuring to recognisable static reference points. Accurate mapping of each feature allows ready re-location of significant sites for more detailed subsequent survey (English Heritage 2003a). Various devices were trialled and used during the surveys over a period of years, most recently in the Severn, but there is no point in recommending any one specifically here, for they will become obsolete: there will be continuous technological advances and new devices will become available.

Electronics and salt water do not necessarily go well together and so some survey teams have preferred to retain recording by pencil on paper or plastic films, which is reliable, if often messy. Wessex Archaeology pioneered the use of a Husky hand-held PC and a laptop to take digital SMR records into the field to guide survey and to make new records directly using PocketGIS software linked to the dGPS. The new records remained 'stand-alone' until the new datasets could be transferred to the SMR periodically in the office base (NK3, 3–4 and 8). Again, recommending any currently available device or software for future work is pointless. Digital voice recorders with a headset microphone were found to be useful during survey in the Severn. Very few recordings subsequently proved to be impossible to understand, even those made in windy weather. Photographic site recording is now universally based on digital cameras. Limited cleaning or augering

Figure 2.8
Ellen Heppell records part of a submerged forest at Thames Site 2, Purfleet, Essex. The GPS comprises a Trimble AgGPS122 – beacon receiver (the antenna and backpack), a differential GPS receiver. The receiver beacon was linked to a Husky Fex21 rugged hand-held computer to log data in the field using PocketFastmap software. In the office data was transferred into a project GIS (in ArcView 3.1). (Image courtesy of and © Essex County Council)

around sites is often necessary in order to characterise sites or place them in context, but excavation does not form part of the RCZAS.

In the North-West (NW4) key heritage assets were placed in order of priority for further work, using a scoring system based on Threat, Condition, Significance, Potential, Rarity and Potential for Designation. This provides a helpful tool for determining future work in the region, in relation to FCERM.

Analytical techniques

In the 1997 publication there was a substantial focus on relative sea-level rise, and this was necessary, for this is ultimately the driving factor in coastal change (Fulford *et al* 1997, 25–49). Since then, however, far more information has become available on past coastal change around England, and offshore from the present land masses, and this is more relevant to the historic environment than relative sea-level change in itself. There was also a comprehensive discussion of the application of Environmental Archaeology to the coastal zone (ibid, 56–73). This original chapter remains very useful, and there is no need to reiterate it fully here. More recent studies built on it will be outlined below. However, a brief review of some aspects of scientific techniques used on the coast, and offshore, is still needed, to provide a basis for understanding the factual basis on which our understanding of coastal change relies.

Geophysical and geotechnical methods

The use of terrestrial geophysical survey techniques to complement field surveys is not a requirement of Phase 2 RCZAS, although it may follow on during subsequent work. In general, geophysical survey has not been widely applied to intertidal sites, or indeed to waterlogged sites of any type, for only Ground Penetrating Radar (GPR) appears to be effective in detecting peat with a low mineral content. Bog oaks and possibly large wooden structures might be detectable (English Heritage 2008a). Measurement of conductivity/resistivity has, however, proved to be successful for detection of major buried features, such as the large Pleistocene palaeochannel at Happisburgh, Norfolk (M Bates, pers comm). Magnetometry may help to define locations of wrecks with ferrous components.

One advantage of studying intertidal sites is that erosion frequently exposes extensive sections in which the relationships of deposits are plainly visible. However, following initial phases of survey, augering or coring to examine and sample deeper deposits may be needed (Fig 2.9). For example, augering was used extensively in Essex to define palaeo-channels and buried land surfaces, and in Langstone Harbour transects were undertaken across the study area to define stratigraphy (Wilkinson and Murphy 1995; Allen and Gardiner 2000, 47–53). Geotechnical investigations (mainly coring) from commercial and other sources frequently provide helpful contextual information.

Offshore, between 2002 and 2011, funding provided by the Marine Aggregates Levy Sustainability Fund (MALSF) and by offshore developers (including the Aggregates and Offshore Renewables industries) has led to an expansion of research into long-term changes in sea level and coastal morphology. This has largely related to changes during the Palaeolithic and Mesolithic (Bicket 2011; Gaffney *et al* 2007; Ings and Murphy 2011). The offshore Regional Environmental Characterisations (RECs) have provided useful information on submerged prehistoric land surfaces and wrecks (www. alsf-mepf.org.uk). New methodologies for offshore investigation have also been applied to far more recent sites, including wrecks (Hamel 2011) and submerged or eroded historic sites, for example the

Figure 2.9
Jack Russell of Wessex Archaeology prepares to sub-sample a core collected from Area 240 off the East Anglian coast for palaeoecological analysis.

medieval town of Dunwich, Suffolk (Sear *et al* 2011). The latter study employed an integrated approach to using historical, archaeological, marine geophysical and diver-based survey of the site. This is not the place to consider in detail the new techniques and their applications, though all have undergone significant advances over the last fifteen years. As detailed in the above publications, they include geophysical prospection methods (sidescan sonar, multi-beam sonar, 3-D seismic survey, magnetometry, sub-bottom profiling), diver-based survey, remotely operated vehicles, geoarchaeology (including collection of samples by vibrocoring and grab-sampling), environmental archaeology (analysis of remains of sub-fossil organisms to provide data on past environments), and scientific dating (including radiocarbon dating and optically stimulated luminescence (OSL) (Sear *et al* 2011). In the relatively shallow water around the Isles of Scilly (up to about 20m) sidescan sonar and sub-bottom profiling, linked to a GPS system, was used to detect submerged peats, which were then recorded and sampled for scientific dating and environmental studies by divers (Cornwall Council 2012, 23–72); while in the eastern Solent creeks draining into the former Solent River were defined using offshore geophysics and near-shore augering and coring (Tomalin *et al* 2012, 88–133).

Artefacts

Besides the durable artefacts common to any archaeological sites – such as ceramics and lithics – waterlogged coastal sediments commonly include wooden structures and other organic artefacts sometimes including leather, textiles, cordage and basketry, bone and antler and even horn. Guidelines to the recording, sampling and conservation of these materials are given in English Heritage (2010c; 2012c).

Environmental archaeology

Environmental archaeology is 'the study of past human economy and environment using earth and life sciences' (English Heritage 2011a). It overlaps with geoarchaeology, which is concerned more specifically with soils (especially ancient soils, known as palaeosols) and sediments, and makes use of chemical and physical analyses, soil micromorphology,

mineralogy and particle size analyses (English Heritage 2007). A key technique in environmental archaeology is the recovery of remains of organisms and their identification. Categories of material examined include vertebrate remains, insects, molluscs, ostracods and other crustaceans, macroscopic plant remains (such as seeds, wood and mosses), microscopic plant remains (pollen, spores and phytoliths), and unicellular organisms including diatoms, foraminifers and testate amoebae. From the known modern habitats of these organisms inferences can be drawn about local environmental conditions from micro- or macro-fossil assemblages. Similarly, assemblages of, for example, fish bones and mollusc shells can provide information on the human exploitation of marine resources (Figs 2.10–2.11). The types of remains preserved depend on the preservation conditions of the deposits being sampled. Acidic conditions in sediments do not permit preservation of bone or shell. In dry deposits most types of organic remains are biodegraded, so that a submerged deposit formerly on land may *now* be waterlogged, but includes only pollen, bone and charred seeds owing to degradation before submergence.

Figure 2.10
Faunal remains from Norwich. Part of an edible crab cheliped (Cancer pagurus) *indicating a commercial crab fishery from the late medieval period (see far left); the remaining shells are of Macoma and other small bivalves which probably reached the town accidentally on the gear of fishing boasts exploiting intertidal shell-fisheries.*

Figure 2.11
St Martin-at-Palace Plain, Norwich. This block of early medieval refuse deposit includes articulated skeletons of herring (Clupea harengus). *Fish bones are more commonly extracted by bulk sieving. (Image by M Sharp, courtesy and © Norfolk Museums Service)*

5 cm

In permanently waterlogged anoxic sediments such as peats and muddy deposits, a wide range of remains may be preserved. However, localised variation in the burial environment can result in unexpectedly good or poor preservation. To obtain material for analysis, deposits must be sampled in the field. This may involve hand-collection, of wood or large bones, for example. More frequently, sample collection may involve coring or augering (vibrocoring offshore), collection of intact monoliths of deposit in five-sided metal tins, arrays of small samples for specific analyses, or bulk samples for extraction of large items. Subsequently the samples obtained must be stored in conditions appropriate to prevent deterioration (usually cooled, excluding light and air). Processing to extract material then follows: a range of physical and chemical techniques is used, to extract the type of material under analysis. Then the items extracted must be identified and the overall composition of the assemblages used to model the local environment at the time of deposition (English Heritage 2011a).

An example of bulk sampling and analysis is provided by the site on the edge of salt-marsh at Leigh Beck, Canvey Island (Wilkinson and Murphy 1995, 187–93). The deposits here comprised clay-based deposits of Roman and 12th–13th century date and were rich in shell, bone and, at some levels, briquetage (salt production waste), probably emplaced to raise ground levels. Biological remains were extracted from bulk samples and quantified. The mollusc shells included food waste, mainly shells of mussel, oyster and cockle, but shells of inedible species and abraded and bored shell fragments showed that these were not simply shell middens. Shelly debris was reaching the deposits by wave transport, or deliberate incorporation from nearby *cheniers* (natural intertidal shell deposits), to raise the ground level (Fig 2.12). Mature intact shells of *Scrobicularia plana* probably represented waste from bait preparation. Fish bones likewise may have included some naturally transported material, but the medieval layers included bones of sharks and rays, herring, conger eel, cod, haddock, whiting, horse mackerel, grey mullet and flatfish at high concentrations. Other bones were of sheep/goat, cattle, pig, small mammals, chicken, dunlin and frog/toad. Charred cereal remains predominated in the Roman deposits, but

Figure 2.12
An interpretation of Roman and medieval shell assemblage, formation from Canvey Island Essex (Wilkinson and Murphy 1995, fig 118). (Reproduced courtesy of and © Essex County Council)

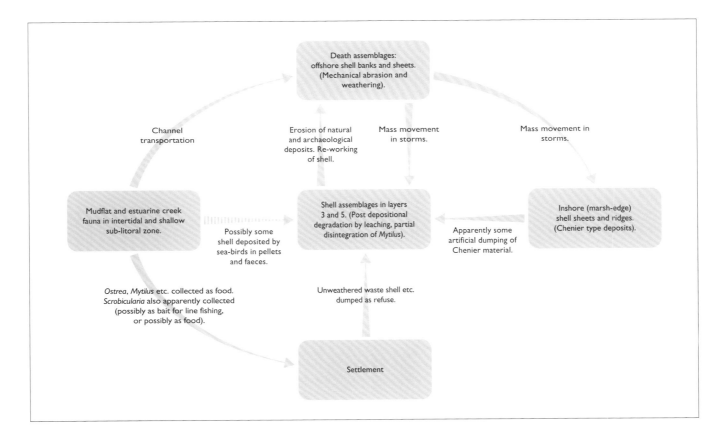

were though to have been imported for use as fuel, rather than indicating local production. Conditions were not ideal for microfossil preservation, but sparse assemblages of diatoms including soil species were present, implying that the site was not regularly inundated. Overall, the results indicated that salt production was of significance in the Roman period, with some food waste disposal, whereas in the 12th and 13th centuries the site appears to have been a fish-processing base. Although the quantities of mammal bone recovered were small, the predominance of sheep/goat is consistent with historic use of salt-marsh and grazing marsh for sheep.

Microfossils, and some types of macrofossils, are generally obtained by sampling sequentially through cores (or vibrocores at sea) or from vertical columns of spot samples through sediments. Combined with samples for scientific dating (see below) this permits reconstruction of changes in the ecology (palaeoecology) of locations through time. Diatoms are single-celled siliceous green algae that have specific habitat requirements in terms of salinity, dissolved oxygen and pH, the type of sediment and local vegetation. Foraminifera are unicellular protists, confined to marine environments, which likewise are proxy indicators of salinity and elevation in relation to mean sea level. Ostracods are crustaceans with calcified carapaces, living in aquatic habitats and likewise provide data on the salinity, current velocity, sediment type, water depth and temperature of a depositional environment. Palynomorphs (pollen, spores and other air-borne particles) are mainly produced by flowering plants, ferns, and mosses, providing data on the character of vegetation around a sampling location. Samples may also include shells of molluscs, remains of insects (mainly beetles), seeds, wood and charcoal, all derived from taxa with specific habitat needs. In practice, what survives in a core or set of samples is related to the local depositional environment. Collection of a series of cores and analysis of micro- and macro-fossils from a range of locations along a coastline can reveal marine transgressions, for example on the Solent shore of the Isle of Wight, where woodland dominated by lime (*Tilia*) was covered by coarse sand and then salt-marsh sediments from around 6000–5000 BP, before becoming fully intertidal (Tomalin *et al* 2012, 26–87).

The vibrocore illustrated in Figure 2.9 was collected from Aggregate Extraction Area 240, off Great Yarmouth, from which Palaeolithic artefacts had been extracted during dredging (*see* Chapter 1 and Wessex Archaeology 2010). Vibrocore 2, from the area where the artefacts were believed to have come, produced only pollen assemblages, dominated by pine, sedges and grasses, together with micro-charcoal particles, and spores derived from much earlier deposits. This helped to place the artefacts within a specific type of habitat, on a floodplain with pine and herbaceous vegetation in the vicinity.

Dating

Scientific dating of intertidal deposits and structures is often essential. A few of the more useful techniques for investigation of coastal sites are outlined below. Occasionally sites and structures may be dated by associated artefacts but this is uncommon: coastal activities often would not have involved bringing datable artefacts to marginal environments. Wooden and timber structures may show some features suggesting a date, for example the differing tool marks and cuts on the wood, defining the type of tool used (Fig 2.13). The species used may also be helpful: the use of softwoods often implies a relatively recent date. However, wooden structures are often of simple form, so a prehistoric oak structure can look identical in the field to a post-medieval example. Consequently, wood, peat or other organic samples were collected and archived during field survey, with a view to selection of key samples for scientific dating later.

Radiocarbon dating on the coast was first used generally to date Sea Level Indicators: contacts in a sediment sequence where a shift from freshwater sediments, such as peats, to minerogenic muds and clays formed in an intertidal environment, indicating a Transgressive Overlap – a shift to marine conditions; or, *vice versa*, pointing to a Regressive Overlap (Fulford *et al* 1997, 25–49). Subsequently it has been more widely used to date archaeological material. Radiocarbon dating depends upon determination of the content of ^{14}C in a sample of organic material. CO_2 is taken up by plants during photosynthesis, and these may then be eaten by animals. Both release ^{14}C back into the atmosphere by respiration. The amount of ^{14}C in the tissues

Figure 2.13
Near Hythe, Hampshire. A typical intertidal structure. Dating can sometimes be achieved from early maps, construction methods or the timber used, but usually scientific dating is required to characterise structures of this type fully.

of living organisms is thus the same as in atmospheric CO_2 at any given time. They are in equilibrium with it. When organisms die, they cease to exchange CO_2 with the air, and the proportion of ^{14}C in their bodies becomes fixed. Their hard parts (mainly wood, shell or bone) may survive degradation by fungi and bacteria and become buried; and the ^{14}C that these parts contain continues to undergo radioactive decay. Measurements on samples of durable biological material, recovered during excavations, indicate the proportion of ^{14}C surviving and, hence the age of the sample. The half-life of the isotope is 5730 years. The determination and its standard deviations then have to be calibrated, to correct for past variation in ^{14}C levels, using computer programs such as OxCal, which incorporate standard calibration curves. The end product is a calibrated date in calendar years (Cal BC/AD), with new error terms.

More recent advances in radiocarbon dating methodology include:

- AMS dating (accelerator mass spectrometry), which permits the use of very small samples (5–10mg). In this method the number of ^{14}C atoms in the sample is measured directly, rather than detecting radioactive decay products. This method not only permits dating of such small items as individual charred cereal grains, but also allows very fine discrimination when taking samples from sediment sequences.

- The use of mathematical modelling to enhance precision of calibration (Bayliss 1998). Modelling makes use of Bayesian statistics, in which prior knowledge of the position of a sample in a sequence can be incorporated.

- Improvements in measurement precision, allowing high precision dates to be obtained.

Dendrochronology depends on the fact that tree species growing in a temperate climate show seasonal variation in growth, following an inactive period in winter (English Heritage nd). This is most apparent in ring-porous woods, such as oak, elm and ash, in which vessels produced in spring are large whereas those produced later in the season are small, so that annual rings are well defined. Ring widths depend on environmental variables such as annual variations in temperature and rainfall. The technique depends upon comparison of ring sequences from sections of wood samples from different sources. Each has its own 'signature' composed of successive narrow and wide rings. The sequences are measured under an optical microscope, by visual inspection or by the use of scaling devices and automatic recorders. A recent advance is the use of X-ray densitometry, whereby X-radiographs of wood sections are scanned by a beam of light and a photocell, measuring density of the wood (hence registering differential vessel sizes). The measurements obtained are compared with master chronologies of known date. Some site sequences may show high correlation with master chronologies, which are nevertheless not statistically significant. This problem can be resolved by high-precision radiocarbon dating of samples to test the correlation, often using Bayesian statistics. By this means, the Bronze Age timber circle on the beach at Holme-next-the-Sea, Norfolk, has been dated to 2050–2049 BC (Bayliss *et al* 1999) (Fig 2.14).

Optically Stimulated Luminescence (OSL) dating depends upon light 're-setting the clock', or releasing energy accumulated by unstable isotopes, in quartz grains suspended in water. Once these grains are deposited as sediment and buried, energy accumulates within crystal lattices by decay of radioactive isotopes in the sediment. Light emitted when a sample is re-heated in the laboratory provides a measure of the time that has elapsed since deposition (English Heritage 2008d). The technique is widely employed to date mineral sediments, especially from material collected from cores offshore, for example off Great Yarmouth, where sands, silts and clays were assigned to successive phases of the Pleistocene, including the Ipswichian (MIS 5e) (Wessex Archaeology 2007). Off the Isles of Scilly, OSL dating of sands associated with submerged peats was undertaken alongside radiocarbon dating. Consistent results were obtained. (Cornwall Council 2012, 111–33). It has also been employed for dating coastal sand dunes as, for example, at Gwithian, Cornwall, where work proceeds.

Palaeomagnetic dating depends on measurements of declination, inclination and intensity of natural remanent magnetism (NRM) in sediment samples. It provides a measure of the geomagnetic field existing when the sediment was formed. From sediments spanning the last 10,000 years relatively high-precision dates can be established by comparison with known pattern of

Figure 2.14
Holme-next-the-Sea, Norfolk. A post of the Bronze Age timber circuit being raised for dendrochronological dating.

short-term variations in the earth's field. Over very long periods of time, the geomagnetic field reverses: the poles change positions through 180°. Such polarity reversals are detectable in rocks and sediments. The present orientation of field is known as normal polarity, the opposite being reversed polarity. Palaeomagnetic dates on sediments at the pre-modern human site on the beach at Happisburgh, Norfolk, displayed a reversed polarity, which helped to constrain the site chronologically to before 0.78 million years ago (Fig 2.15).

Figure 2.15
Happisburgh, Norfolk. Contorted laminated silts overlying gravels.

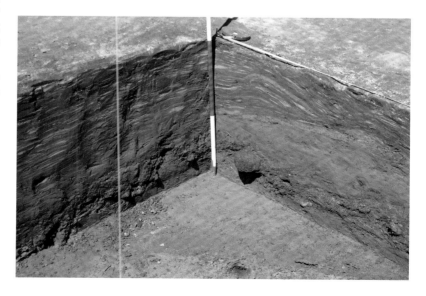

The age of this site is further constrained by biostratigraphic evidence – the presence of tree pollen of *Tsuga* (hemlock) and *Ostrya*-type (hop-hornbeam type), which are unknown in northern Europe after the early Pleistocene, and by the presence of mammalian species, notably 'advanced' forms of *Microtus* (voles), which suggest a date between 0.99 and 0.78 million years ago (Parfitt *et al* 2010).

Health and safety

Coasts are hazardous. Besides the dangers inherent in any archaeological fieldwork, some of the main risks on coasts include:

- getting stuck in soft intertidal mud or sand;
- loss of orientation on featureless tidal flats;
- loss of the sense of time in a bare and blank tidal flat environment, most significantly in relation to tidal cycles, which could result in being trapped or drowning;
- loss of control or power failure of boats or people falling overboard;
- injuries sustained by falls from cliffs, or by leg injuries from tripping into narrow creeks on salt-marsh, or bait-diggers' holes on mud- and sand-flats;
- dehydration, exhaustion, sunstroke and exposure;
- exposure to biohazards, particularly sewage effluent, but also discarded hypodermics in urban areas, with a risk of contracting Weil's disease, Tetanus and Hepatitis A;
- risks from infected cuts and insect bites;
- danger from unexploded munitions where foreshores have been used by the military as ranges;
- risk of gunshot wounds from wildfowlers;
- loss of digital communications; and
- risk of injuries from attacks by birds at nesting sites, or by bulls on unfenced grazing marsh.

Finally, the surveyors themselves could cause damage or disturbance to the environment, for example by disturbing feeding or nesting birds.

The RCZAS programme, involving very many people around the country, resulted in very few incidents and no injuries or deaths, thankfully, but that was because the health and safety of workers was given the highest priority during development of project designs. This *must* be adhered to in future projects.

Each project had its own individual Health and Safety statement, related to local conditions and means of access, but they shared some common features. Reference was made to tide tables before starting work in the intertidal zone. Team leaders were familiar with their use, and also their limitations. There was an agreed protocol for the teams to report to the local Coastguard or Harbour Master and to their own base staff when arriving at, and departing from, sites. A minimum team of two was on site at all times, and safety equipment including first-aid box, compass, rope, water, food and flares was carried. There was a first-aider in each team, and one member of the team had particular responsibility for H&S. Where biohazards were suspected, surgical gloves and antiseptic wash were provided. Special care was taken when entering unfamiliar areas on foot, and team members maintained visual and audible contact at all times. Clothing and protective gear was appropriate to the circumstances. Lifejackets are essential on boats and hard hats were required for cliff survey. People generally preferred to wear the minimum possible during summer survey, though waterproofs and warm clothing were available. Boats were used judiciously, and operated by adequately trained and experienced staff. Small vessels used during survey complied with *'The safety of small workboats and pilot boats – a code of practice'* (Maritime and Coastguard Agency, nd). The teams were provided with VHF radio. The VHF requires certification for use and must be licensed. Mobile telephones must **never** be relied upon during survey since they may not receive a signal behind cliff faces – even very low salt-marsh edge scarps. In the event of an emergency, signals on the local emergency channel would have been picked up by all VHF users in the area, and the Coastguard would have been able to take bearings on VHF signals. Close liaison with managers of designated wildlife sites was always needed to avoid sensitive areas or undertaking survey at times that might interfere with birds or other vulnerable creatures.

During the North Kent survey, three incidents did occur, none of them serious (NK5, 16–17). In one case an engine failure and a fouled anchor on the workboat necessitated cutting away the anchor, and a return to shore, with delays until a new anchor could be obtained. On Burntwick Island a member of the

team became stuck to the waist in soft mud on a rising tide, but was extracted by a colleague standing on firm ground, using throw-lines and rope. Finally one member of the team fell into the Swale Channel from the workboat, but the lifejacket inflated effectively and the person was rescued. Incidental injuries, completely unrelated to project work, have resulted in logistical problems. For example, the project lead for the Severn RCZAS injured his back picking up a toddler at home, necessitating some rescheduling and reorganisation of the project; but such an accident could occur with any project. Accidents happen, but the consequences can be minimised sensibly.

Excavation and monitoring

Survey in the 1980s and 1990s often led fairly directly towards a programme of excavation, sometimes on as large a scale as possible, employing many of the analytical techniques outlined above. For example, survey of the Essex coast was followed by excavation of a Neolithic site and later wooden features at The Stumble in the Blackwater Estuary, Essex (Wilkinson *et al* 2012) (Fig 2.16). In other contexts, full excavation of threatened sites, such as the Bronze Age timber circle at Holme-next-the-Sea, Norfolk – 'Seahenge' – (Brennand and Taylor 2003) was also seen as the best approach at that time. At the latter site large components of the timber structure were also lifted for display at the King's Lynn Museum. This made the results from the site more

publicly accessible, but it imposed major costs to ensure conservation of the timber lifted. These excavations were useful exercises in terms of development of techniques for inter-tidal excavation, besides providing more widely applicable archaeological information and displayable objects, discussed further below.

Excavation and the post-excavation process, however, are expensive. Resources in the 21st century are now much more limited. Following these 'set-piece' formal excavations, subsequent work has been focused more on site monitoring at both sites. In effect, erosion 'excavates' the site without human intervention. Repeated visits to monitor this process, with intermittent collection of significant artefacts, provides much of the information that would be obtained in an excavation, but at a far lower cost. At The Stumble, monitoring has continued by means of artefact collection and augering, but has also defined new areas of the surviving submerged Neolithic landscape and later prehistoric features (Heppell 2006). At Holme, new timber structures, some of them prehistoric but also later structures, including Anglo-Saxon fish traps, have been planned, recorded, and wood samples have been collected for radiocarbon dating (Ames and Robertson 2009). There can be little doubt that monitoring eroding structures and sites will be the way forward. Besides, it can involve contributions by volunteer workers from the local area, making an input to the understanding of their own coastal historic environment (*see* Chapter 6).

3

Coastal change

Pleistocene global sea-level changes, referred to as eustatic change, were on a massive scale, related to climate change associated with interglacial to glacial cycles. These resulted from thermal expansion and contraction of the world's ocean volume and melting and re-forming of land and sea-ice. Climatic fluctuations are defined in terms of Marine Isotope Stages (MIS), working backwards from the present warm stage, MIS 1 (Gibbard and Cohen 2008), alongside regionally developed terminology, such as the 'Anglian' for the first glacial stage in England. The maximum ice advance of the latest, Devensian, glacial stage (MIS 2) was around 22,000 years BP, when global sea-levels fell to 110–130m below those of the present. The distributions of hominins, including pre-modern human species and latterly *Homo sapiens*, were affected by these changes, over almost the last million years. Alternately, extensive areas off the modern coast were dry land or sea.

However, sea-level at any given time and place results not just from eustatic sea-level change, but also from uplift or depression of the earth's crust (tectonic and isostatic change). Isostatic change has resulted largely from repeated phases of ice- and seawater-loading, alternating with phases when areas previously under water became dry land, and ice-free, so free of loading.

In general terms, those parts of the crust that were most recently loaded with seawater, or glacier ice – up to 1km thick in places – are still undergoing uplift or 'rebound' once that mass is removed, whereas locations that have not been recently loaded are stable or subsiding. There has also been longer-term regional tectonic uplift, related to movements of the plates that make up the earth's crust. In southern England, for example, regional uplift over the course of the Pleistocene has led to the survival of a raised beach at Boxgrove at +40m OD. The interaction between eustatic and tectonic/isostatic changes results in a relative sea level rise (RSL), unique to each region and varying through time. Besides this, more localised changes in coastline morphology – the development of near-shore barriers such as dunes and spits, for example – have been critical in determining what areas were habitable locally, and which were not.

The palaeogeography in which the earliest pre-modern humans (hominins) lived in England is summarised in Parfitt *et al* (2005). At this time, 0.78–0.99 million years ago, there was a continuous land connection with the continent, with marine embayments in the modern North Sea region and the English Channel (the Manche Embayment). Rivers with a predominantly west–east axis draining the land mass later to become England included the extinct Bytham and Ancaster Rivers draining the Midlands, the Thames (with a formerly more northerly axis) and the Solent River to the south (Fig 3.1). This early pattern of embayments and drainage patterns was completely modified during subsequent glacial stages.

Figure 3.1
The pre-glacial North Sea and English Channel region, with limit of Anglian ice advance. (Image courtesy of Simon Parfitt, Ancient Human Occupation of Britain Project, Natural History Museum)

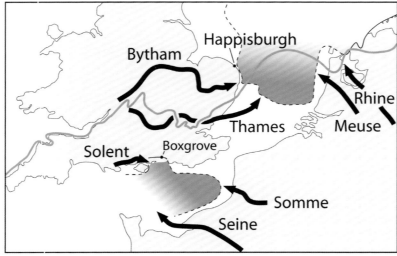

Recent research has provided considerable detail of changing topography on now-submerged landscapes. For example, the Outer Thames Estuary Regional Environmental Characterisation (REC) examined an area of some 3800km² off the coast of Clacton-on-Sea and Dunwich, showing that for parts of the last 500,000 years the area had been dry land, and defining successive submerged river systems related to former courses of the Thames. An apparently early west–east flowing channel extends the early Cromerian course of the Thames/Medway system some 60km eastwards from the modern shoreline near Harwich, while a post-Anglian (post-MIS 11) system is visible further south (Emu and University of Southampton 2009). Archaeological and faunal material has also been recovered from offshore contexts. During 2008, Palaeolithic artefacts – including hand-axes, flakes and cores – were found in gravel at the SBV Flushing wharf, dredged from within Area 240, about 11km off Great Yarmouth. Bones of woolly mammoth, woolly rhino, and other mammals were also found (De Loecker 2011; Wessex Archaeology 2011). The part of Area 240 from which the material came is dominated by two channel features, the oldest of which – Channel A – was probably cut during the late Anglian. The upper fill of Channel A extends to form an extensive floodplain observed throughout the majority of Area 240 and is thought to be of Wolstonian age (MIS 8). It is considered that the majority of the artefacts were originally associated with, and were *in situ* within, these floodplain deposits, whereas the faunal remains are likely to be associated with both earlier and later sediments (Wessex Archaeology 2010). Subsequent studies involving grab sampling and on-board inspection of aggregate dredgings have helped to verify this conclusion. The key point arising from studies at this site is that they demonstrate that, even within a glaciated region, Palaeolithic sites survive offshore, and that they can be placed within a stratigraphic and chronological context.

The submerged landscapes of the English Channel are dominated by a complex series of channel systems, cut into underlying sediments by rivers flowing in a predominantly east–west direction at periods of low sea level during the Quaternary (James *et al* 2011, 53–6). Most channels remain open, but some are infilled with sediment, for example, the former Solent River channel and others north of Cherbourg.

The channel systems are broadly divisible into Northern and Southern Palaeovalleys. Before the chalk ridge between the Isle of Wight and Purbeck was breached after the Ipswichian interglacial, the Solent River flowed eastwards north of the Isle of Wight and turned southwards to discharge into the Northern Palaeovalley of the English Channel (NF1, 18–20; NF4, 29–30). Following breaching, flow was partly diverted westwards. During the Cromerian Complex (MIS 13) the river channel would have been an estuary, with coastal cliffs along Portsdown and east of Arundel, where the Slindon Raised Beach deposits were emplaced. These deposits, at +40m OD, are assumed to have attained this level owing to subsequent tectonic uplift. A comparable situation persisted into OIS 9 or 7, when the Aldingbourne Raised Beach was formed (24–27.5m OD), while the Pagham Raised Beach (c 3m OD) dates to the last interglacial, the Ipswichian (Fig 3.2). By the Devensian low stand, with sea level some 110m below modern sea level, The Solent and adjacent areas would have been fully dry land (Fig 3.3).

Palaeolithic material is most likely to occur offshore within infilled river systems and raised beaches, in the former case mainly unstratified artefacts and faunal remains brought up by dredging or fishermen, but also including the inland *in situ* Palaeolithic site of Boxgrove, Sussex, on the Slindon Raised Beach (Roberts and Parfitt 2000). Extensive areas of the English Channel would have been land surfaces during

Figure 3.2
Pagham, West Sussex. Raised Beach at around 3m OD, c 100,000 BP.

Figure 3.3
Development of the Solent
River system. (© Wessex
Archaeology, © SeaZone
Solutions Limited, 2012 and
© Crown Copyright, 2012.
All rights reserved)

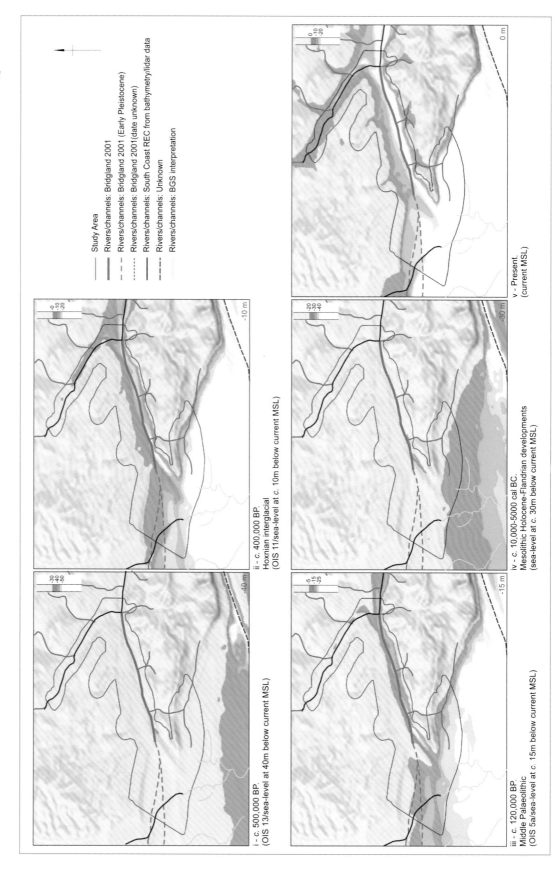

parts of the Cromerian complex, forming a broad low plain with lakes, pools and rivers – all potentially inhabitable. However, during the Anglian glacial stage cold conditions might have inhibited human presence, despite the sea-level low-stand. During the Hoxnian, sea level was around 10–15m lower than today, so that much of the Solent and the Sussex coastal plain was available for human activity (Figs 3.4–3.5). Sites such as Priory Bay on the Isle of Wight may date from around this period (James *et al* 2011, 120–32). East Dorset's rivers were also related to the Solent River, comprising an upstream section of the Frome, Piddle/Trent, Stour and Avon flowing along a watershed incised into Tertiary sediments and separated from the sea by an east–west chalk ridge, which may have breached not later than 70,000 BP (D2, 5–6). Upper Palaeolithic sites include that at Hengistbury Head, Dorset, overlooking the coastal plain.

Pleistocene sediments of the Severn are principally submerged or deeply buried beneath later sediments (S1, 59–68). There are Ipswichian raised beaches at Weston-super-Mare and other locations, while the Burtle Sands Beds were deposited in shoals and sand flats. In Cornwall the Fal Estuary originated in the Quaternary, when it was incised into the Pliocene platform at +45m OD. The valley was submerged during the Ipswichian, but in the Devensian a cliff was created about 3km offshore at –42m OD. The cliffs and raised beaches of the Ipswichian were partly buried beneath head and loess. Meltwater streams later transported tin-bearing sediment into the upper estuary, though with little deposition in the outer estuary (Ratcliffe 1997, 12).

Since the end of the latest (Devensian) glacial stage (MIS 2), around 12,000 years ago, sea-level rise has not proceeded at a constant rate. Besides the underlying RSL rise there were sudden and catastrophic changes, most notably the so-called '8.2 kiloyear event'. This resulted from a vast discharge of meltwater from the northern Canadian (Laurentide) ice-sheet – probably when ice barriers to the north failed, allowing pro-glacial lakes to drain suddenly into the North Atlantic. It led to abrupt global sea-level rise, besides causing climate cooling. At Rotterdam there is evidence for a jump in local sea level of around 2m in 200 years (in addition to long-term RSL rise of perhaps around 1.9m), beginning around 8450 cal BP (Hijma and Cohen 2010). On a shorter

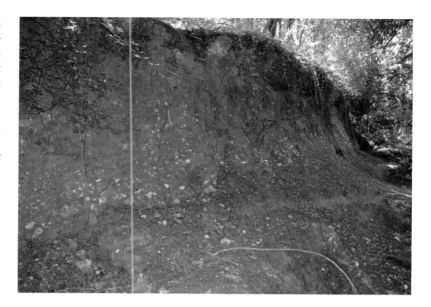

timescale, around 8100 cal BP, a series of submarine landslides off the Norwegian coast caused a tsunami, the second Storegga Slide Tsunami, which deposited sandy sediments up to 80km inland and 4m above present tide levels in Scotland, and probably also had effects on the Northumberland coast (Weninger *et al* 2008). There is no doubt that this event must

Figure 3.4
Priory Bay, Isle of Wight. Artefacts from this cliff-top site were redistributed to the beach below.

Figure 3.5
Medmerry, West Sussex. Recording a large slate erratic, transported by an iceberg along the English Channel during a glacial stage.

Figure 3.6
Sowley Marsh, Hampshire.
Coastal change during
the Mesolithic would have
produced transitional
landscapes looking
comparable to this.

also have had major effects on Mesolithic populations living on the North Sea lowlands, very probably causing deaths by drowning, and affecting the productivity of the coastal wetlands that provided their subsistence base (Fig 3.6).

The submerged Mesolithic landscapes of the North Sea were first termed *Doggerland* by Coles (1998). Subsequently, analysis of 3-D seismic data has produced a detailed image of submerged Mesolithic landscapes across much of the southern North Sea (Gaffney *et al* 2007). Extensive areas of lakes, wetlands, salt marshes, intertidal zones, salt domes and at least ten major estuaries were defined, with some 1600km of streams (Fig 3.7). A major lake and then estuary occurred in the area of the Outer Silver Pit, and was associated with sandbank features. The features mapped do not represent a single period, but subsequent ground-truthing and scientific dating shows that the majority are of Mesolithic date, with underlying Pleistocene features, including gravel outwash plains and tunnel valleys.

By around 8000 cal BP most of Doggerland was submerged beneath the rising RSL of the North Sea and direct land connection between Britain and the continent had been severed,

although there remains the intriguing possibility – unproved as yet – that islands may have persisted in the North Sea as late as the Neolithic, aiding colonisation at that time. The south-western part of the North Sea landscape, lying off the present Essex coast, comprised a lowland area of soft geology, drained by the former seawards extensions of the Rivers Stour, Blackwater, Crouch, Thames and their tributaries. The Thames floodplain overlies a very deep sequence of sediments, consisting of freshwater peats and estuarine clay/silts, over 15m thick at Tilbury. Analysis and dating of these sediments enabled Devoy (1979) to define the process of relative sea-level change from around 8170 years ago. He defined a sequence of transgressive overlaps, named Thames I–V (when the sea advanced inland and clays/silts were deposited), and regressive overlaps (Tilbury I–V), when there was a retreat of marine conditions, and freshwater peats were formed. Sediment sequences related to coastal change from the Mesolithic onwards remain exposed within the intertidal zone or are deeply buried within modern estuarine sediment sequences. Of particular archaeological significance is the Thames III transgression, which resulted in widespread

Holocene fluvial systems
Sand banks (Holocene)
Early Holocene lakes
Relative topography
High
Moderate
Moderate/low
Intermediate /low
Low

0   50
kilometres   N

Figure 3.7
General map of all recorded
Holocene landscape features
in the south-west study
area of the North Sea,
including general
topographic interpretation.
(Reproduced by permission
of Visual and Spatial
Technology Centre,
University of Birmingham)

submergence of a fringe of coastal land, and sites on it, around 3850–2800 years ago.

In Essex, extensive scatters of Mesolithic artefacts occur at River Crouch Site 4, at Rettendon, where Fenn Creek meets the Crouch, and at Maylandsea, on the Blackwater. The Fenn Creek site has been known since the early 20th century and still generates numerous lithics, implying occupation over a continuous period. Here, the land surface is overlain by peat, dated to 3760±70 and 3660±80 cal BP. Consequently, the former land surface remained in a terrestrial context and was reworked for millennia after Mesolithic occupation, so that no contemporaneous surfaces survive, only scatters of lithics. Despite this reworking, augering and boreholes indicate a former east–west flowing channel immediately to the north of the scatter, with potential for organic preservation. The Maylandsea sites are less well exposed, though the lithics appeared to be eroding from a land surface, on which a tree stool was rooted: dated to 4190±80 BP (Wilkinson and Murphy 1995, 62–70). All sites were fully terrestrial when occupied. The Fenn Creek site was revisited in 2001 and erosion was monitored: the salt-marsh edge had retreated by around 2m, resulting in erosion of the peats (E2, 67).

The Crouch and Blackwater estuaries and open coast of Essex preserve extensive areas of later submerged land surface, sometimes with stumps of oak trees rooted in it. For example, at Blackwater Site 8 at Tollesbury a tree stump was dated to 4030±80 BP (ie the surface was submerged after the tree died, but before it could decompose) and numerous lithics and sherds were recovered. The site appears to have become submerged in the early Bronze Age. One effect of this transgression was to seal abandoned dry-land settlement sites beneath layers of sediments, including estuarine clays, detritus mud and peats. Though submerged, at least at high tide, these sites have been protected from the processes of weathering, root and animal disturbance, and truncation by ploughing, that have so badly damaged similar early prehistoric sites that have remained on land. Only now, as the sediment cover erodes away, do former landscapes and sites become visible once more. The results of survey work in Essex during the 1980s at first suggested that similar extensive exposures of submerged prehistoric landscapes, with associated Mesolithic and Neolithic sites, would be found at many other locations elsewhere in the eastern and southern coasts, but it now seems that such exposures are uncommon (Wilkinson and Murphy 1995). Apart from a site on Hoo Flats in the Medway, where a scatter of lithics and pottery of mid- to late Neolithic date on a submerged land surface has been reported, extensive exposures have not been reported (NK5, 1 1). The Essex sites are more significant than was first realised.

To the south, coastal change and prehistoric land use took a different form. Despite the cold phases at the end of MIS 2 (including the Younger Dryas), the climate subsequently warmed, leading to sea-level rise. Mesolithic activity in the Solent, Langstone Harbour

and Chichester Harbour is indicated by lithic scatters. In Langstone Harbour a low, predominantly grass-covered, plain with isolated stands of hazel, birch, pine, oak and elm existed, part of a wider coastal environment extending from Southampton Water to Brighton (James et al 2011, 120–32). Only one fully submerged Mesolithic site has been investigated, at Bouldnor Cliff in the Solent, where lithics and wooden artefacts have been recorded eroding from a former land surface at –11m OD (Momber et al 2011). Poole Harbour did not achieve its present form until around 8,000 BC, when the chalk ridge between Ballard Down and The Needles was finally completely eroded (Dyer and Darvill 2010, 18–19).

The bathymetric and seismic techniques used to study the North Sea, as noted above, have also been applied to west coast submerged palaeolandscapes (Ings and Murphy 2011). The Liverpool Bay would have been a very different landscape in the upper Palaeolithic, one of land and including a series of rivers draining from the uplands into large shallow lakes. In the immediate post-glacial period this landscape would have been cold and dry, probably with some surviving ice, and with tundra vegetation. Humans were present, at least during part of this time, inhabiting caves on the surrounding uplands. During the Mesolithic this landscape was progressively submerged, leading to the formation of an extensive intertidal zone, but with the main rivers persisting, as seawards extensions of the modern Mersey, Dee and other rivers. The drier parts of the landscape would have been covered in woodland, but the river systems would have provided access to the coastal plains, where aurochs, deer and wild boar, waterfowl and fish were available.

Coastal change inferred from terrestrial sediment sequences in the North-West has been reviewed by Barlow and Shennan in NW1, 26–49. Between the Anglo-Welsh border and Formby Point RSL change is dominated by eustatic change, with local geography determining specific coastlines. The Mersey may have been beyond the southern limit of post-Late Glacial Maximum isostatic uplift. On the Wirral at Newton Carr sea level was at –8.97m below present at 6814–6460 BC. Between Formby Point and the River Wyre eustatic change dominates but proximity of the area to the former Lake District and Irish Sea glaciers means that isostatic adjustment is occurring at around 0.1mm/yr. Further north, towards Walney Island, a date on a Sea Level Index (SLI) Point from a peat at Heysham indicates a sea level –20.87m below present at 9096–7962 BC. Isostatic uplift may have been around 0.69mm/yr since 4000 BP. North of this, towards St Bee's Head and on to the Scottish border there are fewer sediments including SLIs, though isostatic uplift has been estimated at 0.95mm/yr between Walney and St Bee's Head. At Black Dubb north of Allonby a sand dune system is dated to 8203–7053, suggesting an ancient coast close to the modern one. Uplift may have been around 0.87mm/yr in northern Cumbria. It is suggested that a highstand 4m above present sea level may have pertained in Cumbria in the early Holocene, with subsequent net negative tendency with transgressive and regressive overlaps.

A glacio-hydro-isostatic model indicates that Ireland was connected at its south-east corner to the main land-mass of Britain by a low sill (up to about 5m at 18,000 BP) of emergent sea-floor, though the connection is thought to have been 'tenuous' since meltwater would have been dammed by the ice mass to the north up to the height of the sill. Defining the final separation of Ireland is thus problematic, though it occurred before around 10,000 BP (Lambeck *et al* 1995).

Relative sea-level rise in the Severn in the South-West led to submergence, with Lundy Island persisting for a while as a peninsula in the expanding Bristol Channel around 12,000 to 10,500 cal BP, but later being isolated as an island (Figs 3.8–3.9). Along with other islands and large rocks in the Severn, Steep Holm would have provided an elevated view across the Mesolithic and Neolithic landscape of the valley (S1, 43).

The coast of the South-West had taken up very approximately its present form by around 7500 BP. Holocene sediments show an initial phase of early silt-dominated deposits, formed in mudflats and salt-marshes, a mid-Holocene sequence of intercalated silts and peats, and a return to silt deposition in the late Holocene. However, there is much local variation and palaeogeographic change was influenced by local variations in topography and coastal morphology, for example the presence of sand banks. Between about 8500 and 5000 cal BC there was rapid sea-level rise, initially 5–6 mm/yr or even up to 7.5mm. From about 5000 cal BC the rate dropped to around 2mm, allowing

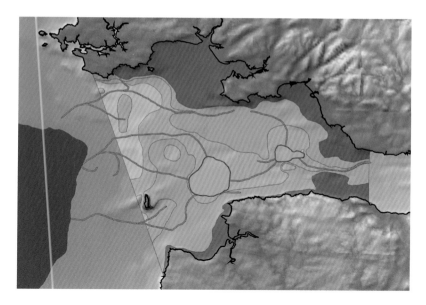

organic sedimentation to exceed sea-level rise with widespread peat formation. Again, local topography affected when and where this occurred in tributary valleys. Around 1500 cal BC there was a shift from the middle to upper Somerset Levels Formation with clastic sediment deposition. Rising sea level caused flooding of the raised bog in the central Brue Valley, associated with the Meare Heath and Tinney's tracks, dated 1550–1450 BC (S1, 59–68).

The Isles of Scilly comprise an archipelago of islands and rocks of metamorphosed Palaeozoic and intrusive granitic rocks about 45km west of Land's End (Fig 3.10). They illustrate, in

Figure 3.8
The Severn Estuary at around 12,000 BP, showing river channels and lakes. (Reproduced by permission of the Visual and Spatial Technology Centre, University of Birmingham)

Figure 3.9
The Severn Estuary at around 10,500 BP. Lundy Island at the head of a peninsula. (Reproduced by permission of the Visual and Spatial Technology Centre, University of Birmingham)

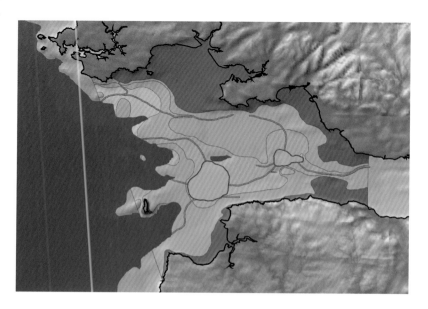

Figure 3.10
From St Helen's, Isles of
Scilly. The numerous islets
are residues from a larger
land mass.

microcosm, the process of progressive submergence that has affected the entire British Isles since MIS 2. At around 7000 cal BC, Scilly was a single island but, by around 4000 cal BC, rising RSL had separated St Agnes, Annet and the Western Rocks from the northern islands. By 2500 cal BC there was tidal flooding between the islands and during the early to middle Bronze Age (2500–1500 cal BC) there was rapid and large-scale submergence, resulting in something approaching the modern pattern of islands. It is suggested that extensive construction of entrance graves at this time may have been part of the population's response to the rapid loss of land (IoS; Cornwall Council 2012, 181–3; Ratcliffe and Straker 1996). Although a small collection of possibly Mesolithic artefacts has been recovered from St Martins, it seems that the islands were not regularly occupied until the Neolithic, and that permanent occupation only began in the Bronze Age. Pollen and archaeological evidence indicates that large-scale farming began at that time. Prehistoric stone structures exposed in cliff sections or in foreshore exposures include hut circles, field boundaries, cairns, cist-graves and stone alignments, but in many cases these are not closely dated. The stone field walls may have originated in the Bronze Age, but probably continued in use as late as the early medieval

period (Ratcliffe and Sharpe 1991). Although most are assumed to be of late Bronze Age date, the chronological evidence is poor. By the early to middle Iron Age, the woodlands of the island had been largely cleared (Scaife 1984). Settlement sites at Halangy Down, St Mary's and Nornour indicate continued occupation between the Iron Age and Romano-British period.

After prehistory

Reconstruction of historic coastlines presents difficulties, for in many areas the evidence has been eroded away or deeply covered by later sediments (Fig 3.11). Data providing a broad overview of coastline change in prehistory are usually good enough for archaeological purposes, but they are not adequate for the last two millennia, since we are often interested in the details of coastal change at particular places. Greater precision is often required and this calls for detailed and costly studies.

In Norfolk some reconstruction of the Roman period coastline is possible (Murphy 2005a), although Roman roads heading for the coast end suddenly at the modern shore with no indication of their original, now-lost, destinations. This is typical of many parts of the lowland areas of the country. During the Iron

Age a major marine incursion into the fens resulted in deposition of silty sediments known as the Terrington Beds which, during the 1st century AD, were exposed as a sub-aerial surface, on which farms were established. Numerous salterns are known from the area. To the east, in the Breydon Water area close to Great Yarmouth, an extensive estuary developed once former coastal barriers had been removed. Estuarine conditions extended to within 7km of *Venta Icenorum* and the entrance to the estuary was guarded by two forts, at Caister-on-Sea and Burgh Castle. To the north of these forts an island later known as Flegg was isolated by the channel of the Bure and a now-infilled channel north of Winterton. Concentrations of beach finds occur in the area between Sheringham and Cromer. In 2003 ditches, a wooden structure, buried topsoil, building materials and pottery were recorded at Winterton, possibly representing a saltern site (N1, 26, 31).

Sea defences, land-claim and erosion

The sea defences of the lowland coasts of England are the archaeological evidence for landscape transformation – the conversion of mudflats and salt-marsh to agricultural land (Rippon 2000, 47–50). Their antiquity is often not recognised, but they dwarf more celebrated military earthworks. They represent an enormous effort, spread over millennia, to enlarge and increase agricultural production,

yet they are widely ignored. Many early sea-banks are still there at their original locations and still serve their original function, though they were often later raised and armoured with rock rubble or concrete blocks. Others, such as the late Saxon Fenland sea-bank, are now well inland, having been superseded by later defences further seawards (Fig 3.12). Some banks were damaged during storms and never repaired. For example, from the 1280s to the mid-15th century there were frequent breaches of flood banks, for example along the Thames. Some breaches were not repaired, partly because of the higher cost of labour after the Black Death. In places land lost over that

Figure 3.11
Covehithe, Suffolk.
A coastal landscape and road truncated by cliff erosion. There is no 'cove' and no 'hithe' at this location today.

Figure 3.12
Heacham, Norfolk.
The Fenland sea-bank, used in World War II to disguise military defences.

period was not reclaimed until the 16th century (Galloway 2009). New reclamation and the maintenance of existing flood banks has therefore not been a continuous sequential process. At times environmental and social factors made them uneconomic. At several locations around Oldbury in the Severn, set-back seems to have occurred in the early 17th century, leading to deposition of intertidal sediments on abandoned fields (S1, 34). In many places, more recent lost reclamations can still be seen, for example fragmentary sections of sea wall breached in a storm-surge of 1897, along estuaries in Essex (Wilkinson and Murphy 1995, 208–9). Despite the builders' aspirations, there was never any assurance that claimed land would *remain* claimed, and this helps to provide a historic context for current programmes of coastal Managed Realignment. Realignment is not new (*see* Chapter 6).

Sea walls of Roman date are known from the Severn Estuary (Allen and Fulford 1990) and suspected in other areas. By the Roman period, in the Severn extensive salt-marshes extended well inland at many locations, with the coast seawards of the modern coast in a few places (S1, fig 27). Reclamation appears to have been undertaken by the military in south-east Wales in the 2nd century and by villa estates on the English shore during the 3rd. However, the evidence for reclamation at this time, other than around the Severn Estuary, is slight and the dating often doubtful. In general there may simply not have been sufficient land-hunger to justify the effort of reclamation during the Roman period. Many areas of coastal wetland were left unclaimed, being used for salt production and grazing (Rippon 2000, 136–7). There is good evidence for a late Roman transgression provided, for example, by the Huntspill Cut 3rd–4th century Roman saltern or the major east–west Roman road, the Fen Causeway, in Norfolk, both of which were later covered by intertidal silt. The abandonment of coastal marshes between the 3rd and 5th centuries AD on the English coast is seen by Rippon (2000, 142–3) as resulting from a range of factors. Marine transgression, economic change, political insecurity, and large-scale population movements were probably all involved. The next major phase of land-claim was in the mid- to Late Saxon period, when parts of Fenland, Romney Marsh and coast of Somerset were embanked. Charters suggest that embankment may have begun in the

8th century in North Kent, but it was certainly underway in many areas by the 10th–11th centuries.

From the 12th century onwards, the general trend was towards reclamation in back-fens and coastal marshes for mixed agriculture, though with local and regional specialisations in production. There was a long-term trend in the high medieval period towards landscape transformation, which can be explained in terms of socio-economic factors: in particular, the evaluation of the costs and benefits of reclamation by large estate-holders – lordly, monastic and episcopal. Nevertheless, independent initiatives by local communities remained important in some areas. Frequently, small parcels of land were reclaimed piecemeal and progressively, but the process is consequently not at all well documented. In the 13th century the establishment of the Commissioners of Sewers was an early move towards, ultimately, national regulation of flood defence and land drainage.

Various methods of embankment were used, depending on local conditions. Sea walls and flood banks were commonly constructed from clay, often dug from a 'back-ditch' or 'borrow-dyke', which itself had a role in land drainage, receiving freshwater from field-ditches and channelling it towards sluices that, in the medieval period, were simple wooden flap-sluices. 'Warping' – placing hurdles, brushwood or stone on the foreshore to trap sediment – was sometimes employed. Some sea-banks were simple earthen structures but others included internal timber frameworks, stone revetments or projecting groynes. Hulks were often incorporated into sea defences as reinforcement.

There was marked variation in forms of land-claim around the country. In East Yorkshire drainage and embankment of low-lying coastal areas in the 17th century is marked by dykes and sluices (cloughs), as at Easington. Sea-banks of this period include Long Bank, Easington, and the Humber Bank. In the Humber Estuary salt-marsh formed behind Spurn Point from the 16th century onwards, especially in the area of Sunk Island, and was progressively reclaimed between 1762 and as late as 1965 (Van de Noort 2004, 160), with more extensive reclamation to the south of Cleethorpes in the 19th century (YL2, 190). Her Majesty's Prison North Sea Camp, originally a Borstal, and subsequently a juvenile prison was established at Freiston on the Wash in

1935. Staff and prisoners were actively involved in land reclamation between 1935 and 1979 (YL4, 40–1). In the area to the west of North Gare, Teeside reclamation took place from the 1860s onwards, preserving a residual dune island from the pre-reclamation landscape (NE5, 77–8).

In Lincolnshire at Tetney medieval saltern mounds were linked by a sea wall – the Old Bank, constructed around 1576 (YL2, 163). Modern settlements lie well inland, and it is thought that large parts of the present coastal zone are a post-medieval reclaimed landscape. Surviving features include defunct sea-banks, sluices and wind-pumps (YL2, 170). Between Donna Nook and Gibraltar Point sea-banks were constructed in the medieval period – including the Crooked Dyke, Sea Dyke and the so-called 'Roman Bank' (YL3, 68). Reclamation continued in the 17th century, with new dykes and sluices, for example Porter's Sluice at North Somercotes, with an associated wind-pump. Sea-banks thought to be of this period have been recorded in a number of parishes, with clay extraction pits at Mablethorpe (YL3, 69). Between Gibraltar Point and the Norfolk county boundary much of the so-called 'Roman Bank' is thought to have been of medieval origin, and a defence certainly was in existence in the 13th century, although the existence of lines of late medieval salterns landwards of it at Wrangle implies that the stretch between Wrangle and Wainfleet, at least, was constructed later, perhaps in the 16th–17th centuries (YL8, 41). It is now up to 2km inland.

The term 'Roman Bank' is obviously misleading, for not only is it of later date, it is also not of one build, but rather a series of banks linked together. New banks were raised through the 17th and 18th centuries (YL4, 65), and reclamation continued into the 20th century. The 1793 sea-bank at Holbeach is approximately 1km inland from the modern shoreline and this continued to be partly functional until after World War II (YL4, 52). At the mouths of the major rivers, such as the Welland, 'training walls' were built to create navigable cuts, besides reclaiming land to either side (YL4, 66). The overall effect was to connect small channels that originally fed into creeks, canalising them into a small number of drains that were controlled by sluices where they cross sea-banks, as at Anderby Creek and Saltfleet Haven (YL7, 51).

The sea-bank in Norfolk, running from Clenchwarton to West Lynn, is considered to be largely late Saxon and medieval. As in Lincolnshire, it often links saltern mounds. Marshland was embanked from the medieval period onwards, partly under the influence of a Carmelite Friary, though with early works funded by the canons of Walsingham Priory. Some banks on the Burnham marshes may be medieval, but there was a renewed phase of embankment in the 17th century. The latest main phase of embankment, completed in 1822 and designed by Thomas Telford, cuts across earlier banks. The evidence indicates repeated phases of enclosure, with new banks enclosing progressively larger areas (N2, 143–6). During ground survey banks were recorded around the Wash and north Norfolk, with timber structures perhaps representing internal revetments of banks, at Burnham Norton and Wells-next-the-Sea. Related features included sluices and channels (N1, 151) (Albone et al 2007).

In Suffolk approximately 200km of sea walls were recorded from aerial photography. The earliest documented reclamation was undertaken around Orford by 1169/70, though aerial photographs also indicate drainage features around the site of Leiston Old Abbey, founded in 1182. At Orford, relict sea-banks inland of modern defences were mapped, corresponding to earthworks shown on an early 17th-century map (S4, 78–87; Williamson 2005, 27). The rate of land-claim in Suffolk was reduced by the late 17th century. There appears to have been a situation of diminishing returns, for the most easily inned land had already been claimed, and the agricultural economy was generally depressed. However, by about 1750 and still more so during the Napoleonic Wars, the combined effects of increasing population and high agricultural prices made further reclamation economic. There were successive improvements to sluices, drains and banks throughout the early 19th century.

Reclamation in Essex was partly related to providing grazing marshes for sheep. 'Wic' place-names refer to dairy farms. The total flock in 1086 is estimated at 18,000. However, unclaimed salt-marsh could also be grazed, so long as refuges and sheep-bridges over creeks were supplied to enable stock to retreat at high tides and in storms. The evidence suggests that land-claim began in the 12th century, with subsequent major phases in the late 16th–19th centuries (Wilkinson and Murphy 1995, 20–210). Parts of Foulness were reclaimed by

1271 to comprise a series of reclaimed marsh areas, independently embanked, separated by channels or 'guts', which filled at high tide. Subsequent inning occurred in the 15th century and excavation of a sea wall has provided a dendrochronological date of 1484–9. Inning was renewed in the 17th century and was completed by 1833. The new land was then increasingly used for arable, with a population peak in 1870, declining thereafter (E1, 5; E5, 3). In the medieval period Canvey comprised five islands, reduced to three by the 16th century. However, the defences deteriorated and the island was repeatedly flooded, being used mainly for sheep. In 1622 Sir Henry Appleton and Joas Croppenburg agreed to renew reclamation, with a new sea wall, and by the 19th century land-claim was completed. Drainage and loss of life in 1953 led to raising of the sea walls and emplacement of flood barriers in the main creeks (E1, 8; E5, 4).

In North Kent sea walls, frequently now well inland, were recorded from aerial photographs (NK1, 48). Burntwick Island and Slayhills Marsh show linear drainage features and embankments of probable post-medieval date, now eroding (NK4, 11). Flood defences of presumed medieval date in East Kent were focused in the Stour valley and around Sandwich (SE1, 47), while a complex history of embankment beginning in the early medieval period is known from Romney Marsh (Rippon 2000, 157–67). Haystack stances occur adjacent to Wantsum Channel, Romney Marsh and near Bexhill (SE3, 63).

Post-medieval drainage and embankment features between Littlehampton and Arundel along the valley of the River Arun are well defined, notably at Climping, associated with water-meadow features or ridge and furrow (SE2, 52 and 59). Between Bosham and Chidham a large breakwater was constructed in the early 19th century, with a view to reclaiming the upper channel. It failed but the bank and timber posts are still visible (SE2, 55). The sea defences for Portsea Island are of medieval and later date (SE1, 16). In Dorset early reclamation is indicated by a reference to reclaimed land at Weymouth of 1242–3. An attempt to reclaim The Fleet took place in the 1630s, while extensive reclamation around Christchurch Harbour is indicated by 1st and 2nd edition OS maps (D2, 14).

There is extensive evidence for land-claim in the Severn, extending back to the Roman period but more often originating in the Middle Ages: there were sea walls at Tidenham in the 13th century (S1, 23). At Lydney Level a series of defunct sea walls reflect progressive reclamation and erosion from the early medieval period onwards: ridge and furrow lies on sediments formed probably in the second millennium AD (S1, 25). The tip of Awre comprises reclaimed land, possibly before the 12th century, but the final Main Sea Wall was of 19th-century date (S1, 26). The Great Wall of Elmore at Bridgmacote has been proposed as a Roman sea wall, largely on the basis of reclaimed land levels, but reconsideration suggests that reclamation was of late medieval date at the earliest (S1, 28; S4, 108–11). Surface finds from Longney have been taken to imply Roman and early medieval reclamation, though this is disputed. The latest reclamation there is no later than 1327. At Arlingham medieval ridge and furrow lies behind the earliest reclamation, with subsequent land-claim in the 18th and 19th centuries (S1, 28–9). Elsewhere in the Severn, the remains of reclamations involving cribbing or faggoting are recorded – using bundles of brushwood held in place by stakes. Groynes and the walled cliff defence to protect the grotto at St Audrie's Bay have been recorded (S4, 55) (Fig 3.13). At Steart Point westwards migration of a shingle spit resulted in marsh accretion landwards of it and progressive stages of medieval reclamation can be distinguished from embankments and ditches: during excavations as part of the Managed Realignment Scheme in 2012 medieval moated settlement enclosures have been defined. Further to the east, a 2nd–4th century Roman farmstead site could imply an earlier phase of embankment (Ed Wilson, pers comm). In the North-West, sea defences on the west of Walney Island, pre-dating 1537 and probably raised at the instigation of the monks of Furness Abbey, comprise a 1km long earth bank (NW1, 165).

Between the 13th and 16th centuries there was extensive coastal erosion, resulting in the loss of at least 173 coastal settlements around England (Cracknell 2005, 267–9). There were numerous losses of villages in Holderness and the Lincolnshire coast. The medieval settlement of Skegness had an adult population of 140, a haven and castle, but was destroyed in a storm of 1526. It was subsequently rebuilt about a mile inland, with 14 households in 1563, declining to 10 by the early 18th century. It

Figure 3.13
St Audrie's Bay, Somerset.
The South-West RCZAS field
team examines a collapsing
walled cliff defence.

remained a very minor settlement until later development of the seaside resort (YL3, 54). In Holderness erosion continued into the 19th century, resulting in the loss of villages at Great Cowden and Owthorne, lost by 1844 (YL2, 190). Several Norfolk villages are known to have been lost, including Ness (near Caister-on-Sea), Eccles, Little Waxham and Shipden (N1, 11). Leiston Abbey, Suffolk, was relocated inland in the 14th century (Fig 3.14), owing to repeated flooding, while at Skinburness, Cumbria, a bank some 2km long and up to 8m wide was constructed as the port itself was eroded in the 14th century (NW1, 195). The best-known loss, however, was of Dunwich, Suffolk, where there was continued erosion of the town, mainly from the 14th century onwards (http://www.dunwich.org.uk/). In the North-West the World War II defences of Walney Island, including a Heavy Anti-Aircraft Battery, have been subject to erosion, as shown by aerial photography from 1946, 1950 and 1980, while the medieval motte and bailey at Arlingham is severely eroded (NW3, 36–7).

Occasionally construction itself has had effects on coastal stability. The World War II six-inch gun emplacements of the emergency coastal battery at Hunstanton were fired only twice, raising doubts about the stability of cliffs, extensively dug into to provide chambers and tunnels. They were not re-fired, lest all collapse (N2, 215). During the Holocene mining and China Clay extraction continued to cause

Figure 3.14
Leiston Abbey, Suffolk.
An early example of
relocation. The first
abbey was constructed at
Minsmere on a site that
became unsustainable
owing to flooding. The
rebuilt abbey of the
1360s includes re-erected
12th-century columns.

Figure 3.15
Lostwithiel, Cornwall.
The medieval bridge is
partly choked by sediment
derived from upland tin
streaming and other sources.

deposition of sediment and changes in the position of the tidal limits in Cornish estuaries (Fig 3.15), especially in the 18th and 19th centuries (Ratcliffe 1997, 12).

For the more recent historical period, resources such as historic maps and charts, landscape paintings, watercolours, drawings, engravings, photographs, postcards and written accounts are valuable for recon-struction of coastal change (McInnes 2011). They are of variable quality and reliability. Some early maps, for example, are topo-graphically inaccurate, emphasising features thought by the compiler to be especially significant, and minimising others, with an

overall distorting effect. Some painters – including those working in the picturesque style, including Turner, were concerned with composition and dramatic effect, not necessarily literal accuracy of depiction. By contrast, artists belonging to, or influenced by, the Pre-Raphaelite school 'sought to depict nature in very exact manner', so their works are as factually reliable as photographic images. Topographical descriptions may also be valuable. Used judiciously these sources permit detailed reconstruction of changes such as beach depletion, land-slips and flooding in the last few centuries.

4

The coastal historic environment

Since 1997 – largely from the RCZAS and other English Heritage projects, though also from projects undertaken using other funding sources – a great deal has been learnt. It is not possible to present and discuss all the new results here. Instead highlights, some interesting but lesser known sites, and especially sites threatened with erosion, will be emphasised. Inevitably, this will lead to omission of some well-known and even iconic coastal historic assets. For a fuller discussion *see* Murphy (2009).

Palaeolithic

The enormous changes in coastlines since the earliest known biologically pre-modern human activity around *c* 0.78–0.99 million years ago,

and also methods used to investigate offshore Palaeolithic sites, have been considered briefly in Chapters 1 and 2. Both this very early date and the new methodologies were unimagined in 1997. The evidence for an early human presence before the Anglian (MIS 12) was slight and sometimes disputed at that time.

Results from the sites on the modern shore at Happisburgh, Norfolk, demonstrate that hominids inhabited an upper estuarine part of the early Thames towards the end of an interglacial, within a wider habitat of boreal forest of pine and spruce (Fig 4.1). Habitats in this vicinity would have included the tidal river, salt-marshes, freshwater marshes and a grass-dominated floodplain grazed by herbivores. The richness of food resources provided by this range of habitats may partly explain hominid

Figure 4.1
Happisburgh, Norfolk.
Excavation on the beach
in 2011.

presence, but the fact remains that cultural adaptations to a cooler climate would have been necessary – potentially including the use of fire, shelters and clothes (Parfitt *et al* 2010). This was not the only early hominid presence at these latitudes, for later lower Palaeolithic activity has been demonstrated, around 700,000 years ago, though in a phase of warmer Mediterranean-like climate, on a former floodplain at Pakefield, Suffolk (Parfitt *et al* 2005).

Elsewhere, stratified lower Palaeolithic sites are not common on the coast. Some other sections of the Norfolk coast, especially between Cromer and Happisburgh, are composed of Pleistocene till overlying, in some cases, pre-glacial land surfaces and palaeochannels, which have the potential to produce more early sites. Lithics and faunal remains, derived from these formations, have been recorded (usually unstratified) on beaches, but it was the notable concentration of lithics in the Happisburgh area that led to the investigation of the sites there (Wymer 2005). Similarly, the site of Priory Bay on the Isle of Wight was first indicated by loose lithics on the beach, eroding from gravels at the top of the nearby cliff (IoW, 27). Field evaluation of this cliff-top exposure demonstrated that some artefacts are derived but others are in mint condition in some gravel units, with potential for *in situ* material being present, although the absence of faunal and palaeoenvironmental evidence placed uncertainties on dating (Wenban-Smith 2003).

Wymer (1999, 179 and fig 43) concludes that the glaciers of the Anglian and Devensian have, with very few exceptions, destroyed any evidence for lower Palaeolithic archaeology in the north of England. Nevertheless, there are isolated hand-axes from Redcar and Blackhall rocks in the North-East (NE1, 63). Further south, the Holderness and the Lincolnshire coasts lie at the western margin of the North Sea basin and Doggerland. At Sewerby a buried Ipswichian cliff line has been reported (Boylan 1967) but the majority of middle–upper Palaeolithic finds from this coastline were found unstratified, comprising lithics and bone projectiles (YL2, 172). In the Lincolnshire Marsh and between Gibraltar Point and the Norfolk county boundary it is probable that extensive areas of deeply buried Palaeolithic land surfaces survive, though artefacts are few and mostly again represent unstratified finds from beaches; for example a lower Palaeolithic

blade from Anderby (YL3, 63). They could have come from offshore. Other finds in Suffolk include hand-axes on the beach at Benacre, with Pleistocene faunal material from Kessingland. These sites suggest that more early sites remain to be discovered, eroding from the cliff bases in north-east Suffolk (S3, 36). On the north bank of Holbrook Bay, at Stutton, lithics include a rolled leaf point, straight-tusked elephant, lion and mammoth bones; and at Harkstead rhino and elephant bones with hand-axes. The material is eroding from low gravel cliff faces (S3, 66). In the Thames and Medway estuaries within the RCZAS study area in Kent, Palaeolithic sites lie within the Boyne Hill terrace (Terrace 2) near Gravesend and in head deposits east from the Medway. Occasional finds of artefacts during fishing and dredging indicate the presence of submerged sites, and there is an unstratified hand-axe from Darenth Creek (NK6, 58).

At the west end of the Bembridge on the Isle of Wight, a raised beach occurs between Forelands Fields and Culver cliff and hand-axes are reported from the beach at Whitecliff Bay (IoW, 35). The lower to early upper Palaeolithic is known from unstratified artefacts from the Solent and its present shorelines (Wessex Archaeology 2005a, NF1, appendix A) with assemblages from gravel pits and other contexts, especially between Southampton and Warsash (NF1, 21–2). There is a concentration of offshore lower Palaeolithic material around Selsey (SE3, 29 and fig 13). Interglacial deposits are uncommon in the Solent, except at Stone Point, where Ipswichian silts and peats are inter-bedded within gravels (NF4, 31). In Dorset Palaeolithic finds are correlated with exposures of Plateau Gravel (D2, 5). Occasional finds of lithics from the Palaeolithic indicate some activity in Cornwall (Wymer 1999, 187) but there is nothing from the Isles of Scilly (IoS, 65). Mammoth tusks and lithics, including hand-axes and one Levallois core, are reported from Watchet and adjacent areas including Doniford and St Audrie's (S1, 40).

Late upper Palaeolithic material has come from only one fully intertidal site, at Titchwell, Norfolk, dated to the very end of the period, *c* 10,00–8000 BC (Wymer and Robins 1994). The site was first detected from lithics found unstratified on the beach, which may have been transported there in eroded blocks of sediment, but this led to attribution of the material to a palaeosol beneath a later peat and minerogenic sediment sequence. Elsewhere around the coast individual artefacts or small groups of them have been recovered. A presumed upper Palaeolithic blade was found within an area of Mesolithic occupation at Withow Gap, Skipsea, East Yorkshire (YL2, 172). A uniserially barbed bone point of probable upper Palaeolithic date came from beneath lacustrine peats, associated with Hornsea Mere (YL2, 75). Upper Palaeolithic material is known from the Solent south east of Needs Ore Point, apparently representing a coherent assemblage (NF4, 31). Recent survey work indicates that the upper Palaeolithic site at Hengistbury Head, Dorset has been almost destroyed by erosion (Cole *et al* 2012). In the North-West the study area would have had a coastline approximately 15m to the west in the late upper Palaeolithic. Lithics have come from Kirkhead Cave, Lindale Low Cave, the rock shelter at Carden Park, Cheshire, and the remains of an aurochs with barbed antler points from Poulton-le-Fylde is dated to 13,500–11,500 cal BC (NW1, 50–1). Most caves inhabited in the upper Palaeolithic have received extensive archaeological attention since the 19th century, so that early excavation has removed most of the deposits they contained (Fig 4.2). However, the contents of now-submerged caves offshore are untouched, though plainly difficult to access. Their potential is considered in Chapter 5.

Figure 4.2
Kents Cavern near Torquay, Devon. Francis Powe, great-grandfather of the current owner, excavating in Kents Cavern in the early 1900s. (Reproduced courtesy of Nick Powe, Kents Cavern)

Concentrations of loose unstratified lithics on beaches, combined with inspection of well-attributed faunal remains in museum collections for evidence of butchery cuts, have given the first indication of nearby stratified eroding sites (Fig 4.3). They are usually far from being *just* 'stray finds'. However,

Figure 4.3
Natural History Museum, London. Simon Parfitt and Patricia Wiltshire discuss Pleistocene faunal remains from Norfolk beaches.

systematic and careful recording of locations is necessary if their full potential is to be realised. In the future increasing coastal erosion is likely to produce more lithics and faunal remains on beaches. Careful observation of the distribution of such material is likely to point towards definition of new stratified sites.

Mesolithic

Since 1997 a significant group of Mesolithic settlement and activity sites has been recorded on the north-east coast and these have modified our perception of settlement patterns. A Mesolithic hut on the coast at Howick Burn, Northumberland, has been dated to c 8000 cal BC and was in use for some 100–300 years (Waddington 2007). A late Mesolithic lithic scatter site on a buried land surface at Low Hauxley is associated with marine mollusc shells and some bone and it is possible that structural remains may survive (NE1, 182; NE5, 164). Peat outcrops on the foreshore at Low Hauxley, Druridge Bay, Amble Bay and Cresswell Ponds range in date from the 6th millennium cal BC to the medieval period (NE5, 273–91). Peats at Low Hauxley, Cresswell and Budle Bay have produced lithics (NE3; NE4, 15) and a peat shelf on the lower shore, at Low Hauxley, intermittently exposed during sand scour, showed human and animal footprints (Fig 4.4), which have been planned in detail (NE6). Two Mesolithic lithic scatters at Ross Links, Budle Bay are being exposed by dune deflation and migration (NE2, 15). Finds from Holy Island include bevelled pebbles, thought to be related to processing of seal skins (NE1, 224). Further large lithic scatter sites have been recorded at Newbiggin-by-Sea, and Lyne Hill

and the scatter at Ness End, Holy Island, composed mainly of debitage with few implements, was inspected during field survey (NE1, 171; NE5, 214). Other lithic scatters, mainly to the north of Hart, include the major Mesolithic sites of Crimdon Dene and Filpoke Beacon, the latter dated to 8760±140 BP, with other assemblages from Blackhall. Crimdon Dene is close to the shoreline and the Blackhall sites are on the cliff edge (NE1, 84; NE2, 36) and so vulnerable to erosion. Field survey at Crimdon Dene unfortunately did not relocate the lithic scatters reported in the early 20th century, and organic horizons within dune systems proved to be of recent date, although the possible location of the site was defined (NE5, 95–6). Mesolithic scatters have been recorded further north on the coast between Easington and South Shields. (NE1, 126–7) with a barbed antler or bone point from Whitburn beach. It might, however, have been derived from offshore peat beds (NE1, 135). Intertidal peats and other sediments at Hartlepool Bay have produced lithics of Mesolithic date (Waughman 2005).

Besides producing lithic scatters, the Holderness coastline is of significance since here meres and sediments associated with wetland habitats of the type that would have characterised extensive areas of Doggerland survive in cliff sections. At Withow Mere, Skipsea, bone projectile points and remains of deer and elk or giant elk have been reported from lake sediments, and lithics are known from the vicinity (YL2, 172–3). The meres along the Holderness coastline are considered to have formed in the Dimlington Stadial (late Devensian, MIS 2), remaining as open water and wetland into the Holocene. Early organic lacustrine deposits at Skipsea Withow Mere have been dated to c 9880 BP (Fig 4.5). It is thought to have been about 1km across by the medieval period, but now has almost entirely been lost by erosion (YL6, 44). Field survey also recorded peats and other sediments associated with Barmston Mere, with undated buried land surfaces and peats at Atwick, Holmpton and Easington, infilled ponds at Easington, Holmpton and Withernsea and a palaeochannel at Easington (YL6, 96).

Although occasional lithics, including tranchet axes, have been reported from locations in the coastal zone of Norfolk, by far the most significant Mesolithic site is on Kelling Heath. This raised area of land would have

Figure 4.4
Low Hauxley, Druridge Bay, Northumberland. Footprints in foreshore peat exposure. (Reproduced courtesy of Archaeological Research Services Ltd)

Figure 4.5
Skipsea Withow Mere,
East Yorkshire. A section of
organic lacustrine deposits
in a cliff face.

Figure 4.6
Cross Sand, off the East
Anglian coast, showing a
raised geological feature,
recorded by marine
geophysical survey.
(Image by Andrew
Winterbottom courtesy
of the United Kingdom
Hydrographic Office,
Maritime and Coastguard
Agency data © Crown
copyright and database
right)

provided a wide view across lower areas of land, from which hunters could view the movement of prey – now submerged beneath the North Sea. The locality has produced large collections of lithics. Offshore a now-submerged rock outcrop at Cross Sand, 160 × 25m (Fig 4.6), standing 46m above adjacent scour, could have had a similar function as a point from which to observe migrating prey (Murphy 2007, 10). At Weybourne a grey clay deposit beneath peat on the beach has produced lithics, faunal remains, human skeletal remains, amber and shale of Mesolithic date (N1, 23).

In Suffolk, most Mesolithic and Neolithic artefactual material has been recovered from ploughsoil. However, sites extending beneath sediments infilling estuaries should, potentially, be better preserved. There are indications of such sites, as at Nacton on the Orwell and on the Stour, while sites on higher ground at Kessingland and Hollesley (at 10m OD) may well extend beneath adjacent marshes (S3, 8). In Essex, the stratigraphic context of extensive lithic scatters have been discussed in Chapter 3, at Crouch site 4 (Rettendon Parish), where Fenn Creek meets the Crouch, and at Maylandsea, on the Blackwater have been described in Chapter 3 (Wilkinson and Murphy

1995, 62–70). Augering and boreholes demonstrate that the Crouch site was in close proximity to a palaeochannel, with potential for organic preservation. At Lower Halstow in the Medway, Kent, a Mesolithic lithic working site still produces a few flints (NK5, 48) and Tankerton Beach, Whitstable, has produced numerous Mesolithic finds (NK1, 63). Early Mesolithic sites, comprising hearths and lithic scatters are concentrated mainly in East

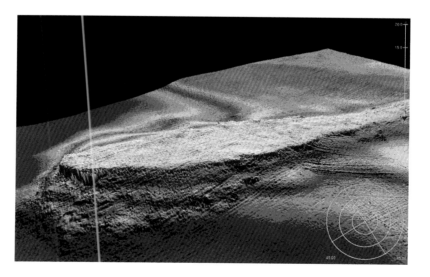

Hampshire and West Sussex, whereas later Mesolithic sites are more widely distributed (SE3, 32).

Between Bouldnor and Hamstead on the shore of the Solent numerous unstratified Mesolithic/Neolithic picks and tranchet axes have been reported, perhaps in part derived from offshore deposits (IoW, 95; Tomalin *et al* 2012, 166–78). From the Solent itself numerous artefacts are known from fishermen's finds (NF4, 34). Only one fully submerged Mesolithic site has been investigated, at Bouldnor Cliff in the Solent (Fig 4.7), where lithics were noted eroding from a former land surface at –11m OD (Momber *et al* 2011). In palaeogeographic terms, the site forms part of the Solent River system, which drained eastwards and then south, ultimately, into the Channel River to the south of the Isle of Wight (ibid, fig 12.5), being fed by tributaries including Southampton Water, and rivers flowing through Langstone Harbour and Chichester Harbour. However, during the post-glacial there never was a channel in the western Solent flowing fully west to east, rather a channel flowing east into the Solent system and south via the Yar. Palynology of the submerged land surface indicates a pine, oak and hazel-dominated woodland in the vicinity with occupation, presumably in more open conditions, on a sandy river bar. Subsequently, back-ponding related to rising sea levels led to peat development, an expansion of grasses and wetland taxa and, finally, salt-marsh development and marine submergence. Overall, occupation in the Western Solent was within a wetland, fed

by the Lymington River in the period 6060–5990 cal BC; after *c* 6000 cal BC brackish water and salt-marsh extended into the area. The western end of the Solent was breached around 4000 years ago, resulting in progressive erosion of the deposits. A peat terrace, at between 11–11.5m OD has been traced for about a kilometre, producing lithics, a single vertebra of pike, worked wood, charcoal, heat-shattered flint and string. The recovery of wood is significant (ibid, 84–7), but its interpretation remains uncertain. It includes worked roundwood, wood chips and larger wood, including a tangentially split piece of oak. Much of the material is charred.

In Dorset the evidence for Mesolithic activity is focused on the Portland/Fleet and Southbourne/Hengistbury headlands, though these geographical features were formed during the most recent marine transgression and are undergoing erosion (Cole *et al* 2012). Poole Harbour, Christchurch Harbour and The Fleet underwent embayment inundation and may be expected to include sediment sequences (D2, 6). Mesolithic material from the Isles of Scilly (IoS, 65) indicates at least occasional visits by sea from the mainland. Lithic scatters of Mesolithic date occur on a palaeochannel at Oldbury-on-Severn. One unit within the peat sequence is dated to 5310 ± 70 BP and represents a soil trampled by cattle, deer and possibly humans, apparently representing the earliest footprints on the English shore of the estuary (S1, 33). This part of the estuary has produced evidence for the burning of reed swamp between 5500–4000 cal BC with occupation continuing until peat formation ceased about 2840–2138 cal BC (S1, 33). At Minehead, foreshore peat and submerged forest exposures include microliths. Charcoal suggests burning of reed peat (S1, 40). Bell (2007) has proposed a model of Mesolithic seasonal movement in which lowland 'aggregation camps' near estuaries were occupied by entire communities over winter, with seasonal dispersal to the open coast and uplands at other times of year.

In the North-West the distribution of Mesolithic material is influenced by the positions of contemporary coastlines, which were inland of the present coast in some locations (for example Eskmeals, Cumbria) but up to 8m below present sea level in others, including Newton Carr, Wirral (NW1, 51). Lithic scatters and middens have been recorded at Hilbre and Walney. The majority of sites are

Figure 4.7
Archaeologists recover palaeo-landscape samples in purpose-built sample boxes from the Mesolithic site of Bouldnor Cliff in the Solent. (Image by Michael Pitts courtesy of the Hampshire and Isle of Wight Trust for Maritime Archaeology © Michael Pitts)

late Mesolithic. Sub-fossil human footprints are recorded from two levels within sediment outcrops at Formby, predominantly of children and females (Fig 4.8): the uppermost footprints are at least as old as 3350–1730 cal BC, the lower examples may well be Mesolithic (Gonzalez and Cowell 2007). There are similar remains at Crosby (NW4). Associated animal footprints included red deer, aurochs, roe deer and wolf. Sites on the Wirral are associated with peat foreshore outcrops, c 5400–5200 cal BC (NW1, 71, 104). To the north a single Mesolithic site was recorded at St Bees (NW1, 185).

Plainly, the Mesolithic of the coast is much better understood than in 1997. Significant finds include footprints on sediment shelves, (at Low Hauxley, Oldbury-on-Severn and Formby), reflecting hunting and gathering activities. The footprints might also point to *destinations*, so that the makers of the prints might be linked to the sites of their homes. The excavation on an eroding cliff-top at Howick Burn shows substantially constructed huts on the coast, occupied for extensive periods. These were not suspected in 1997.

Most finds, however, comprise lithics or bone artefacts from the shore. Some may have been derived from offshore deposits (for example at Whitburn and the Solent shore), but others were certainly stratified within palaeosols and peats, stratified beneath later estuarine or peaty deposits, or under dune sands, or related to palaeochannels. Peat deposits, such as at Bouldnor Cliff, have the potential to produce organic artefacts, including wood, though the logistical problems associated with fully submerged sites limit the areas that can practically be excavated. Elsewhere lithic scatters such as Fenn Creek, Essex, probably extend into a buried palaeochannel, almost certainly including organic sediments and, consequently, preserved artefacts of wood and other organic materials. The inland waterlogged site at Star Carr, Yorkshire, has, up until now, been the one site providing data on this aspect of the Mesolithic, but there are others on the coast, quite probably as important, which need further investigation.

Neolithic

As for previous periods, unstratified beach finds are frequent: for example, Neolithic leaf-shaped arrowheads and other evidence of activity have been reported from two sites on Holy Island and there are isolated records of stone axes from the coast between Low Newton-by-the Sea and the Scottish border (NE1, 205). Flamborough Head has produced a concentration of lithic scatters and occupation sites of Neolithic–Bronze Age date (YL2, 34–5). South from Flamborough Head along the Holderness coastline, unstratified lithics and lithic scatters, including exotic polished axes, are known from a number of coastal locations

Figure 4.8
Formby, Lancashire. Peats and land surfaces occur within the dune system, and beneath it. The palaeosol visible here is considered to be quite recent, probably 20th century.

(YL2, 174–5). The few known Neolithic artefacts from the Lincolnshire coast between Donna Nook and Gibraltar Point and Norfolk are likewise stray finds from beaches (YL3, 64; N1, 13). Around Newtown Harbour on the Isle of Wight, peats, lithic scatters, including a porphyritic rhyolite Neolithic axe, have been recorded (IoW, 97). Neolithic material is generally sparse in the Solent area, but does include lithics (NF1, 23). From the area of Poole Harbour, a jadeite axe from Parkstone originated in Monte Viso above Turin in the Italian Alps. Another axe, from Moordown, originated from Brittany, and a third from Bankes Heath is from the Preseli Hills (Dyer and Darvill 2010, 60–1). The sparse Neolithic material from the Isles of Scilly suggests intermittent, but probably not permanent occupation (IoS, 67). Neolithic lithic scatters and axeheads also occur within sediments of the Severn (S1, 33). Lithic scatters have been reported from the North-West but few are from the present coastal zone. Many scatters in Cumbria occur around the 6m contour line, the level of the high stand of c 3800 BC (NW1, 59). Despite proximity of the Langdale axe factory, few of its products are recorded from near the coast. Lithic scatter sites are recorded from North End Haws and Eskmeals, but at low elevations, vulnerable to erosion (NW1, 158–9).

Earthwork structures and settlement sites also occur in coastal locations, though definite attribution to the Neolithic from aerial photographic data can be problematic. However, where excavation has taken place dating is more secure, an example being the probable causewayed enclosure at South Shields Roman fort (NE1, 128). Long barrows in the coastal zone of the North-East include examples at Lingrow Howe, Street Houses, Loftus (NE4, 20; NE1, 64). Oval and U-shaped enclosures at Overdale Wyke, one only 100m from the present cliff edge, may be of Neolithic date (NE1, 84). At Easington, partly beneath a Bronze Age barrow to the east of the flood defences, excavated post-holes are interpreted as the remains of a rectangular building, associated with hearths and refuse pits and a rich early 4th–late 3rd millennium BC artefact assemblage (YL2, 174–5; Van de Noort 2004). The henge excavated on the shore, also at Easington produced a cremation burial dated to 2500–2000 cal BC (YL2, 177). In Norfolk causewayed enclosures were recorded from

aerial photography, at Roughton and Salthouse. They are small and circular, lacking the multiple ditch systems seen elsewhere in the country (N2, 28–9). The Roughton example, part of a wider complex of monuments, is directly associated with two long barrows and a possible oval barrow. The NMP mapped 22 examples of long barrows, curvilinear forms and mortuary enclosures in the study area (N2, 35), but only one probable cursus, at Hanworth. Few coastal settlement sites are known, although an example from Redgate Hill, Hunstanton, produced inter alia, cereal remains, domestic and wild mammal bones and shellfish (N1, 10; Ashwin 2005; Bradley et al 1993, 61–9). A pit group at Blakeney Freshes on a low rise within coastal marsh produced early carinated bowls through to Beaker sherds with associated lithics (Birks 2003; Jordan 2004). The causewayed enclosure at Freston, Suffolk, is only peripherally in the coastal zone, at an elevation of 30m: it encloses a crop mark of a rectangular timber building, which could be Neolithic, but shows similarities to Anglo-Saxon timber halls (S4, 65–6). Besides this, monumental structures were uncommon in Suffolk, consisting of two oval enclosures at Levington, and a site comprising two concentric circles of pits, at Boyton, overlooking the River Ore (S4, 21–6). Neolithic material from Kessingland is associated with a possible mortuary enclosure. Here and at Benacre Broad the sites overlook valley fens, where there is potential for preservation of environmental evidence (S3, 36). Twenty causewayed ring-ditches have been recorded between Sizewell and Leiston, varying in size from 8–60m across; however, dating of these sites has not been confirmed (S3, 46). There is a focus of prehistoric activity in Hollesley and Boyton on a spur of land overlooking the Ore, comprising a possible late Neolithic/EBA timber circle, causewayed ring ditches, within a probably later field and enclosure system, Bronze Age metalwork finds and, overlooking the Butley River at Boyton a trackway, enclosure and ring ditches (S3, 52). Long barrows also occur in coastal locations further south, for example on Beacon Hill, Rottingdean (SE2, 90).

In the South-East Neolithic settlement sites including cut features, as opposed to artefact scatters, are uncommon, and largely confined to river valleys in Kent and Sussex. A midden including marine shell, lithics and ceramics is recorded from Chark Common, Hampshire.

Monumental structures such as long barrows, cursus and causewayed enclosures occur in the South-East but are under-represented, with very few in Hampshire (SE3, 35).

Neolithic structures in the Solent area include enclosures, a hearth, a possible trackway and a late Neolithic pit at Lepe, dated 2900 BC (NF1, 23; NF4, 36). Near The Needles is a Neolithic mortuary enclosure, with associated Bronze Age barrows and a linear bank (IoW, 75). The Calderstones indicate the presence of one megalithic tomb in the Liverpool area, but most funerary monuments in the North-West are in the uplands (NW1, 52).

Other contexts include coastal sediments, middens and wooden structures. Intertidal peats and other sediments at Hartlepool Bay have produced a range of Neolithic artefacts, including polished axe fragments, wooden artefacts including a lid and a hurdle panel from the early 4th millennium cal BC, tentatively interpreted as part of a fish trap. It was located within a palaeochannel and was associated with stakes. In addition, the site has produced a mid-4th millennium burial, perhaps an early example of a 'bog burial' (Waughman 2005). A Neolithic to early Bronze Age midden on Cowpen Marsh, Teeside produced lithics and faunal remains of domestic species (NE 1, 99). The wooden 'structures' from Withow Gap at Skipsea were, at least in part, accumulations by beavers, since clear toothmarks were observed (McAvoy 1995; YL2, 174–5). However,

other components, including rods and stakes have been interpreted as remains of trackways or platforms (YL2, 177). V-shaped post groups on the beach at Wootton and Quarr on the north coast of the Isle of Wight are interpreted as simple fish traps (Loader *et al* 1997, 12–16). The area has produced abundant lithics. By around 3500 BC peat extended from the shoreline and woodland including oak was established. Wooden trackways were constructed across the peat, extending seawards. Rising sea level led to the woodland being submerged after around 3000 BC. Just inland lithics and late Neolithic pottery were deposited, implying a related settlement (IoW, 69).

The Crouch and Blackwater estuaries and open coast of Essex preserve extensive areas of submerged land surface with Neolithic settlement activity on it, as noted in Chapter 3. One area has been studied intensively before and during the RCZAS: a mudflat known as The Stumble, north of Osea Island in the Blackwater Estuary, Essex (Wilkinson *et al* 2012). At The Stumble there is an extensive intertidal exposure of Neolithic land surface, (approximately 560 × 160m when first located), at an elevation of –0.20 to –0.45m OD. The Neolithic palaeosol survives extensively, and the eroding surface is still littered with flint artefacts, waste flakes, and sherds of pottery (Fig 4.9). In 1985–6, a programme of systematic artefact collection, test-pitting, small-scale area excavation, and palaeo-

Figure 4.9
Lithics and ceramics from the intertidal Neolithic site of The Stumble, Blackwater Estuary, Essex.

environmental analyses was undertaken. Subsequent work has involved monitoring and recording areas of land surface, newly exposed by erosion (E3). During the early to middle Neolithic, from 3685–3385 cal BC, the area was low-lying land, around 1km from the nearest tidal creek. It was drained by freshwater streams. Soils formed on London Clay 'head' supported primary woodland dominated by lime, oak and hazel. There were small-scale woodland clearances associated with farming and exploitation of wild foods and other resources. Abundant charred remains of cereals (mainly emmer wheat) and flax, and some bone fragments of cattle and pig, were recovered, together with charred remains of wild plant foods, (hazelnuts, sloe, hawthorn, roots and tubers), and some bones of red- and roe-deer. Although the occupants of the site were plainly farmers, remains of wild plant foods were just as abundant as those of cultivated crops, indicating that the collection of wild nuts, fruits and roots remained important. As relative sea level rose, the Blackwater Estuary expanded and, by the later Neolithic, soils in the vicinity were becoming waterlogged, freshwater streams were becoming tidal creeks, and a zone of salt-marsh expanded progressively inland. Rising groundwater resulted in the death of trees at the site, (and ultimately preservation of their root systems), and a thin organic sediment spread over the former land surface. The area became increasingly uninhabitable. The latest evidence of human activity on the land surface is from a burnt flint mound, dated to 2490–2285 cal BC. Recording in 2001 showed that there was extensive vertical erosion at the site, exposing larger areas of the Neolithic surface especially to the west. Salt-marsh in the area has retreated by at least 10m since the latest OS mapping (E2, 10).

Neolithic sites that were evidently not settlements are also now exposed in the intertidal zone. At Purfleet, in the Thames Estuary, there is a foreshore outcrop of peat, the base of which is dated to 2554–2313 cal BC, and which overlies a Neolithic land surface. Shells of land molluscs from this surface show that it was exposed long enough for woodland to develop on it, before peat formed in increasingly wet conditions. The site has produced polished stone axes (one of them of Cornish Greenstone), and an unstratified butchered bone of an aurochs but hardly any other lithics and no ceramics. The site was revisited in 2001 and was found to be partly eroding laterally and partly covered with concrete blocks to arrest erosion (E2, 5). Dense charcoal spreads occur on the land surface elsewhere for example at Dovercourt and in the Blackwater Estuary, often mainly of oak, mostly unrelated to any evidence of settlement (Wilkinson *et al* 2012, 86–90). In part these seem to represent a phase of clearance immediately pre-dating submergence in the late Neolithic (4180–3990 BP).

The latest radiocarbon date from a site occupied before submergence of the Essex coastal fringe is from a pit at Jaywick, which produced part of a Beaker (2460–2144 cal BC). Re-inspection in 2001 showed that breakwater construction using granite blocks had resulted in beach accretion which, for the time being, will protect the archaeology beneath (E2, 12).

Neolithic land surfaces and associated settlement features are known from other parts of the country, but they are not common. In North Kent a middle to late Neolithic site producing ceramics and lithics is known from Hoo Flats (NK5, 11and 33). Features include probable pits. Timber trackways and linear features are associated but are likely to be of later date. Current excavations at the Medmerry, Sussex Managed Realignment Scheme, are producing early Neolithic lithics and ceramics in pits and other features dug into brickearth, sealed by alluvium (Jonathan Sygrave, pers comm). On the Wootton/Quarr coastline of the Isle of Wight, Neolithic post alignments and hurdle trackways, with smaller post settings, have been recorded (Tomalin *et al* 2012, 192–201).

A key question for the Neolithic is the extent to which finds and sites from modern coastal locations represent a specific interest in the sea and coastal resources, or whether they are merely the accidental seaward edge of a more extensive terrestrially based settlement pattern. The unstratified finds of lithics (including polished axes) could indicate the proximity of eroding sites or could, for example in the Poole Harbour area, relate to long-distance trade, at least in part by sea. The earthworks and settlement sites do not differ in any obvious way from inland sites and have not produced evidence of consumption of marine foods, except for the marine shells from Redgate Hill, Hunstanton. At Hartlepool Bay and at Wootton/ Quarr, hurdles and V-shaped post structures

Figure 4.10
St Mary's, The Isles of
Scilly, Bant's Carn. Entrance
graves of this type were
constructed extensively
during a phase of major
coastal change in the early
to middle Bronze Age.

have been interpreted as fish traps but there is no supporting evidence for this interpretation. Indeed carbon isotope ratios in the collagen of human bone suggest a shift away from consumption of marine foods in the Neolithic, though the scale of this remains debated (Milner *et al* 2004: Richards 2003). Other Neolithic coastal wetland sites seem to have been related rather to hunting, burial and other activities. The evidence for exploitation of marine resources is therefore slight for this period.

Bronze Age

To a greater extent than the Neolithic, the coastal archaeology of the Bronze Age is dominated by monumentality. Round barrows, cairns and other ritual structures are reported from the coastal zone almost all around the country. Their distribution is summarised in the RCZAS reports and other sources: from the North-East (NE1, 129, 208; NE2, 7), the North York Moors (NE1, 65), the rest of Yorkshire and Lincolnshire (YL1, 91; YL2, 138; YL5, 50), Norfolk (N2, 52–5), Suffolk (S3, 37; S4, 27–30), Essex (Wallis and Waughman 1998, 220), South-East (SE1, 41–4; SE2, 45 and 90), Isle of Wight (Loader *et al* 1997, 16–17; IoW 37 and 79), New Forest (NF1, 3; NF3, 27–9), Poole Harbour (Dyer and Darvill 2010, 61–3), the South-West and Isles of Scilly (Bell 2013;

Ratcliffe 1997, 17; Parkes 2000, 8; IoS, 97; Cornwall Council 2012), and North-West (NW1, 18, 52–3, 130 and 185). As for the Neolithic, however, there is no reason to think that monumentality and the coast were especially linked in most areas. One possible exception may be on the Isles of Scilly where there was a massive loss of land by submergence around 2500–1500 cal BC, coinciding with construction of at least 92 entrance graves (Cornwall Council 2012; Fig 4.10). Unstratified finds from beaches could again relate to nearby settlements or burials, or could represent placed deposits, a consequence of ritual activity (Fig 4.11).

Figure 4.11
Sutton Hoo, Suffolk. Bronze
Age axe found unstratified
on the shoreline during
the Suffolk RCZAS
(by the writer).

Two sites actively eroding on the coast have recently been investigated. The Bronze Age cist cremation cemetery with associated cairns at Low Hauxley, Northumberland, is located beneath actively-eroding sand dunes (Fig 4.12). It has been subject to several phases of recording and excavation, mostly unpublished, but summarised by Clive Waddington in NE3. The cairns were constructed on slightly raised ground with leached brown earth soils (NE3, 10–11), though at lower elevations there was localised waterlogging with peat development, prior to dune formation beginning in the 1st millennium cal BC. Beakers have been recovered from cists, and radiocarbon determinations in the range 2140–1640 cal BC have been obtained (NE3, 29). The part of the cemetery that survives is still buried beneath dune sand and is thus in an undisturbed state, but lies within an area of dune recession.

Along the north coast of Norfolk submergence of back-barrier environments in the later Neolithic led to deposition of intertidal clays and silts over earlier peats, eventually forming a land surface on which the timber circle at Holme-next-the-Sea was constructed in 2049 BC, on dendrochronological evidence. The site, though now exposed on the beach, was originally located when the shoreline was some hundreds of metres to the north (N1, 10:

Ashwin 2005; Brennand and Taylor 2003). A walk-over survey and subsequent monitoring showed that the original site lay within a complex of timber structures, comprising a second circle, platforms, a trackway, post-group and timber-lined pit, besides Anglo-Saxon fish traps (Norfolk Archaeological Unit 2003; Ames and Robertson 2009).

In summary, barrows and other monumental structures that may or may not have contained burials occur widely, and in a range of environments from cliff-top locations (from North Yorkshire to the Isle of Wight) to locations now buried beneath dune systems (Low Hauxley) or in former salt-marsh (Holme-next-the-Sea). In general the cliff-top sites are probably best seen as a continuation of a wider inland ritual landscape, unrelated to the sea. However, monument construction in the then recently formed coastal habitats, as at Holme, may suggest a more specific coastal focus, as may the entrance graves on the Isles of Scilly.

Evidence for settlement, enclosure and agriculture is more sporadic, being revealed as crop marks of presumed later prehistoric land-divisions, excavations at aggregate sites or by direct coastal erosion. A middle Bronze Age to Iron Age occupation site at Barmston, Yorkshire, at the margin of a former mere, dated to around 1500–800 BC, comprising cut features, cobbled

Figure 4.12
Low Hauxley,
Northumberland.
Excavation of a half-eroded
Bronze Age ring-cairn,
July 2013.

surfaces, hearths and ovens was originally interpreted as a 'lake settlement' but in fact was constructed on an earlier peat (YL2, 176–7). The earthwork known as Danes' Dyke, which cuts off Flamborough Head, is not well dated, though associated lithics found in the 19th century have been employed to suggest a Neolithic or Bronze Age date for construction. Evidence of Bronze Age settlement and agriculture on the East Anglian coast is tentative. Enclosures, some with internal round-house ditches, are recorded from several locations, including Letheringsett-with-Glandford, but the dating is unproved. Extensive areas of coaxial field systems are known, including one from Ormesby St Margaret. Some components of the systems have been shown by excavation to include middle Bronze Age pottery; other components are probably later (N2, 62–6). It is probable that here, as elsewhere, middle to late Bronze Age and Iron Age landscape features underpinned later development of field systems and settlement. In Essex, excavation of cropmark complexes in advance of gravel extraction have demonstrated a late Bronze Age date for round-houses, enclosures and watering holes, related to a grazing economy (Wallis and Waughman 1998). Prehistoric crop marks at Fordwich and Chalk Hill in Kent include enclosures, land boundaries, pit groups and other linear features of Bronze Age or later date; a Bronze Age settlement was recorded in the Lydden Valley (SE1, 41–4). The Bronze Age field systems and settlements at Gwithian, Cornwall and Brean Down, Somerset, were both sealed by dune deposits (Megaw et al 1961; Bell 1990). Brean Down represents the most fully investigated site in the region, comprising five phases of activity, from the Beaker period to the late Bronze Age, with evidence for settlement, burial, cultivation, animal husbandry, fishing, fowling and shellfish collection. On the Isles of Scilly permanent occupation began during the Bronze Age (IoS, 67–71). Numerous stone hut circles are recorded, some of which are of this period, typically associated with field systems (Fig 4.13). A site at Roanhead, Cumbria, included a sub-circular structure on a shingle ridge, lithics and a Group VI axe; it is possible that erosion had removed hearths and other features (NW1, 159).

Activity also extended onto coastal wetland areas. Intertidal survey in the Essex estuaries revealed a range of wooden structures, mostly dating from the Bronze and Iron Ages (Wilkinson and Murphy 1995, 132–52). At Fenn Creek, South Woodham Ferrers, a brushwood platform, dated to 2730±60 BP,

Figure 4.13
Porth Cressa, St Mary's, Isles of Scilly. A Bronze Age hut exposed in eroding cliff face.

was associated with two skulls and may have had a ritual function. Other structures comprised brushwood spreads, platforms, hurdles, posts, post-lines and one, late, bridge over a salt-marsh creek (516–390 cal BC). They are dated mainly to the 2nd–1st millennia BC. Functional interpretation is uncertain in most cases, but some are thought to be staithes, landing stages, minor bridges and possibly fish traps. They indicate continued activity on the salt-marshes after these areas had become uninhabitable. At the mouth of Wootton Creek on the Isle of Wight a long timber alignment was built around 800 BC, while at Binstead small rectangular timber structures were emplaced, associated with a timber alignment (Tomalin *et al* 2012, 201–8). At Shinewater Park on the Willingdon Levels a late Bronze Age timber platform, some 2000m³, was constructed on piles driven into peat. The presence of hearths suggested some domestic activity, though human skeletal remains may suggest ritual 'placed' deposition (Greatorex 2003). Bronze Age hoards are recorded on coasts in Kent, East Sussex, West Sussex (especially around Selsey) and Hampshire (notably around Portsmouth and Hayling Island) (SE3, 37). Urnfield cemeteries occur on the coastal plain, for example in Langstone Harbour, and on cliff-top sites (SE3, 39). During excavations at the Medmerry, Sussex Managed Realignment Scheme, middle–late Bronze Age features include round-houses, a well with preserved organics, a burnt mound and a cremation cemetery sealed beneath alluvium (Jonathan Sygrave, pers comm). Excavations for the Royal Edward Dock in the Severn produced human remains and a bronze rapier (S1, 36–7), and middle–later Bronze Age lithics and pottery with associated charcoal and animal bones occur within peat north of Avonmouth. Peat deposits associated with a former mere on Kilnsea beach produced remains of a sewn-plank boat (YL2, 176–7).

Bell (2013) has reviewed the evidence for wetland and dryland Bronze Age activity around the Severn Estuary and has proposed a model of the relationship between the two. Barrows and other burials, deposits of metalwork, ditches and field systems, wells, cattle footprints (especially along the Welsh shore), 'burnt mounds', settlements and intertidal wooden structures have all been recorded. It is thought that the concentration of field systems around the Bristol Avon represents a more intensive form of land use during the later Bronze Age, while the concentration of probable ritual deposits of metalwork in the early, middle and late Bronze Age reflects the Avon's importance as a communication route between the Thames, Kennet and Severn. In general, Bell interprets the evidence as indicating seasonal exploitation of coastal wetlands for grazing, fishing and salt production with more intense use during the middle Bronze Age. Temporary settlements were established in wetlands. Associated permanent settlement sites would have been at the wetland edge (for example Brean Down) or further inland, linked to the wetland areas by drove-ways and trackways.

Boats plainly must have existed before the Bronze Age, but the first direct evidence for sewn-plank sea-going craft, with examples also from the Humber and Dover, and their associated paddles, besides offshore spreads of metalwork at Salcombe, Devon and Landon Bay, Kent, probably marking wreck sites, demonstrate a new focus on the sea (Murphy 2009, 61; Van de Noort 2011). Similarly, the earliest evidence for salt production in England, discussed in more detail below, is of Bronze Age date.

Iron Age

Apart from salterns, discussed separately below, the Iron Age coastal landscape is visible in aerial photographs as field systems, enclosures, promontory forts and settlement sites, some of which have been excavated. In many cases, however, field systems and settlement enclosures are not well dated. Some originated in the late Bronze Age, if not earlier, and some continued in use into the Romano-British period.

Along the coast between Low Newton-by-the Sea and the Scottish border multivallate forts, settlement enclosures and field systems have been recorded (NE1, 208). The fort at Spindlestone Heughs, overlooking Budle Bay, is an example; it comprises an inner enclosure, 100 × 60m with double ramparts and annexes, while the cliffs of the Heugh delimit the south side. Forts at Fenham and Scremerston are also cliff-edge sites, and are eroding. Field inspection at Scremerston did not detect any surviving earthworks (NE5, 192) but at Fenham there were slight earthwork traces (NE5, 203). Multivallate forts and hillforts including

Howick Hill, Fenmanhill and an Iron Age defended site at Westfield, west of Seahouses, were recorded in detail from aerial photography. At Howick earthwork features of an enclosure still survive and at Fenmanhill a second circular enclosure only 93m to the north-west was recorded (NE4, 21). Craster Heugh enclosure is a roughly triangular stone-built structure comprising ramparts and a ditch on a ridge of Whin Sill (NE2, 19). The headland at Tynemouth is thought, from excavated evidence, to have been the site of a promontory fort, as may Dunstanburgh have been (NE1, 66), though at both sites the construction of medieval castles has obscured the archaeology of the earlier phases. Elsewhere in Northumberland early settlement enclosures and other crop marks were recorded from aerial survey. They are generally rectilinear with internal hut circles (NE2, 8). Farmstead enclosures at Cambois and Hauxley were recorded before destruction by industrial activities (NE1, 175), though the enclosure earthworks and a possible hut platform of a small site have recently been recorded at Dunstanburgh (NE1, 175; Oswald *at al* 2006).

Most known Iron Age settlement sites between Whitby and Reighton in Yorkshire are sited between 750 and 1200m inland, as at Cloughton (YL5, 50). However, field survey identified new settlement sites at Cloughton and Reighton, where ditch sections were seen in the cliff (YL5, 87), implying that a more extensive settlement landscape of these periods remains to be discovered. Extensive areas of cropmarks in Holderness are likely to be of Iron Age origin. Iron Age features, ceramics, cremations and occasional coin finds are known from this coastline. Probable promontory forts have been identified at Briel Nook and Gull Nook, Flamborough Head (YL2, 36; YL6, 97). Ditch sections exposed in cliff faces in several parishes between Bempton and Spurn are attributed to the Iron Age (YL6, 97).

Along the Norfolk coast there is low-lying coastal fort at Holkham (Fig 4.14). It is around 260m across and irregular in form. It was originally in salt-marsh below the 5m contour, but later reclamation has left it within grazing marsh, with a covert behind (N2, 71–4). Within the county as a whole enclosures and settlements, besides field systems, appear to be similar in morphology from aerial photographic evidence, whether of Iron Age or Roman date (N2, 81–101). They are typically sub-rectangular, square, and sometimes with internal ring ditches. An example at Heacham in coastal salt-marsh was square, about 50m across, with an entrance, enclosing three ring-ditches of houses 12–17m in diameter. The surrounding area was surrounded by enclosures and trackways. At Snettisham, one partly

Figure 4.14
Holkham, Norfolk.
Earthworks of a low-lying
enclosure. Though now
in grazing marsh, with
plantations behind, this
was originally constructed
in salt-marsh.

excavated enclosure was in use from the 1st–3rd centuries AD. Lying within a very extensive area of enclosures and trackways on the Greensand Ridge, the system includes four square enclosures, each representing an area of settlement or corralling.

A similar situation applies in Suffolk, where settlement and other enclosures, mainly rectilinear, some including round-houses, cannot definitely be assigned to a specific period. As in Norfolk, settlement enclosures were frequently associated with extensive areas of rectilinear field systems. In some cases the layout implies stock management (S4, 35–7). Unenclosed settlements include pits, four-posters and circular houses. Tribal groupings are implied by coin hoards from the beaches – a hoard of Iceni coins came from Covehithe and Trinovantian coins from Alderton (S3, 10). Coins of Dubnovellius type have been found within an area of cropmarks at Shotley, with a dispersed coin hoard within a known trackway. Both sites are less than 300m from the current shoreline (S3, 68). The coins from this locality imply that a site was eroded in the 1970s–1980s. An Iron Age settlement site is known from Burrow Hill, Boyton, at 15m OD in an area that was probably originally coastal. Three large ditched enclosures are known and excavation has produced pits, a round-house and Iron Age pottery (S3, 53).

In the South-East there is a possible Iron Age promontory fort at Hamble Common, delineated by a single rampart earthwork, and another headland, at Wickor Point, Thorney Island, has an enclosing ditch 7m wide and 120m long, and could be a promontory fort (SE2, 48; SE2, 84–5). There are concentrations of Iron Age coins around Broadstairs, Kent, and also around Chichester Harbour. Coins are generally more common near the coast in this part of the country, rarer inland. The fort at Tournerbury on Hayling Island may have been associated with trade (SE3, 40–2).

Iron Age/Roman field systems at Ovingdean and Telscombe are both on high ground in cliff-top locations (SE2, 91). In the New Forest aerial photographic analysis indicated the presence of square and other enclosures at Lower Exbury and a D-shaped enclosure at Inchmery, all potentially of Iron Age date (NF3, 26–7). Hillforts occur at Buckland rings, north of Lymington, Ampress and Lower Exbury, on a promontory. At Efford landfill site round-houses and pits, with evidence for seasonal salt production, were excavated (NF1, 24). Further south-west there are numerous promontory forts (Fig 4.15). On the Helford Estuary, cliff castles/promontory forts, including that at Dennis Head, and hillforts, as at Gear, were centres within a populated landscape of enclosures known as rounds (Reynolds 2000, 7). Roundwood and Pendennis are promontory forts/cliff castles around the Fal (Ratcliffe 1997, 18). Bronze Age settlement and agricultural systems apparently continued, with some introduction of Glastonbury Ware in the 5th century BC, often of gabbroic clay from the

Figure 4.15
Bolt Tail, Devon. Earthworks of a promontory fort.

Lizard. Burials include inhumation cist graves (IoS, 71).

Iron Age to Roman features on the Cumbrian coast, and further south between Seascale and Egremont were cropmark enclosures (NW3, 19–20). Hillforts in the region, some overlooking the coast, for example at Helsby, Cheshire and Swarthy Hill, St Bees, are poorly investigated but some seem to have been abandoned by the later Iron Age. Some ditched enclosures have been investigated, but in general the period is poorly investigated in this region (NW1, 53). Field systems are recorded on the Cumbrian coast at Watch Hill, Nethertown and St Bees (NW1, 160). There is a hillfort at Swarthy Hill, north of Maryport, above Allonby Bay, about 130m N–S. Two ditches have been excavated, but produced no diagnostic pottery. Part of this site has been lost by erosion and further investigation is needed (NW1, 186–7). Wolsey, south of Silloth, includes several Iron Age enclosures with hut circles, and similar enclosures occur at Beckfoot.

These landscape features do not differ in any significant sense from those of Iron Age date inland. The promontory forts and defended sites in salt-marsh or somewhat further inland (for example Warham, Norfolk) merely exploit the topography and may be comparable in function to hillforts elsewhere. However, there have been some excavations of sites directly exploiting coastal wetlands, for example excavations in advance of works for the second Severn Crossing revealed Iron Age settlement at Hallen, comprising structures and post-built round-houses dating to the 2nd–1st century BC. The site was interpreted as a short-lived seasonally occupied site based on sheep and cattle grazing (S1, 34). Coastal manufacturing centres occur on the Ower Peninsula in Poole Harbour, which has produced imported ceramics, dated c 200 BC, with evidence for manufacture of shale artefacts, pottery, salt and metalwork (Dyer and Darvill 2010, 66–7). In North Kent, Iron Age to Romano-British middens, salterns, pottery kilns and cremations occur at many locations in the Medway Estuary and at Cliffe and Shornemead (NK5, appendix 2; NK6, 13–14 and 33).

Burial and ritual sites are known from some regions – the 'Arras' culture of East Yorkshire is characterised by square barrows and 'cart burials': barrows of this form are known from the coastal parishes of Flamborough, Bempton, Bridlington and Barmston (YL2, 177–9). Apparent ditches of square barrows are recorded from the Norfolk coastal zone, typically 8–16m across (N2, 71–4). Norfolk is noted for significant finds of Iron Age metalwork, including torques. However, these finds, mainly from the west of the county, lie inland (N1, 23–6). On Scilly there are three cliff castles: Giant's Castle, St Mary's, Burnt Hill, St Martin's and Shipman Head on Bryher, though whether these are in fact defensive or ritual structures is disputed (IoS, 71).

Later Iron Age trading sites, or emporia, are generally indicated by exceptional assemblages of imported artefacts, and concentrations of coins, both native and continental. Coastal trading settlements are known or suspected from a number of locations. Concentrations of finds of staters (including a single find of 206 staters), with some copper-alloy sheets is reported from the area around Weybourne and at Sheringham and Runton, Norfolk, suggesting locations of trade (N1, 23–6). In the south of England, trading sites have been defined at Hengistbury Head and at Mount Batten, near Plymouth (Cunliffe 1987; 1988). From the late 1st century BC, the focus of trading activity seems to have shifted once more, to Poole Harbour (where a timber and limestone rubble mole, in two sections (c 160 and 55m in length respectively, and c 7.4m in maximum height) has been recorded. It has an upper surface of limestone slabs (Dyer and Darvill 2010, 39–40). Selsey Bill has produced abundant high-value Iron Age coins, including British and continental issues. Associated settlement evidence implies that there may have been a trading site there (Fulford et al 1997, 166). It has been suggested that major Iron Age fortifications, such as Castle Dore, to the west of the Fowey Estuary, and the associated promontory fort on St Catherine's Point, were also involved in redistribution of high-status goods, traded for tin. Finds from Castle Dore include imported glass bracelets, lump glass for bead manufacture and Roman amphorae (Parkes 2000, 25). Tin ingots have been recovered from Bigbury Bay, Devon, apparently from the site of a wreck. They are of an unusual form, shaped like cattle knuckle-bones (astragali), and Diodorus Siculus specifically describes British tin ingots as being of this shape.

Continental trade contacts were not confined to the south of England in the Iron Age, although sites in the south have been more extensively investigated. The site at Meols, on the tip of the Wirral (unfortunately largely destroyed by erosion during the 19th century), appears to have been a beach trading site (Griffiths *et al* 2007). The earliest evidence for long-range trade contact from the site includes three Carthaginian coins of the late 3rd century BC, but 1st century BC–1st century AD Armorican and Roman coins and continental metalwork were also found.

Romano-British

Military sites, salterns and Roman havens are discussed in more detail below. Along the Holderness coastline cropmark sites related to a dispersed pattern of rural settlement – enclosures and field systems – are widespread, though not well dated in general. The shoreline of the Roman period in this area could have lain up to 1–2km east of the present coast, so specifically coastal sites have in general not survived (YL 2, 179–81). Between Gibraltar Point and the Norfolk county boundary only one late Romano-British settlement is known, at Fishtoft – apparently an agricultural settlement unrelated to salt production (YL4, 63). There was some replanning of the agricultural landscape, however: at Hopton-on-Sea, Norfolk, a rectilinear planned field system, 1.7 × 0.9km, apparently of middle to late Roman date, overlies later prehistoric features (N2, 81–101). Villa estates were developed in some areas. There is a probable villa at Sutton, Suffolk, of 'corridor house' type; surface finds suggest use in the 1st and 4th

centuries, with associated enclosures and field systems. A second possible villa site is at Wherstead overlooking the Orwell; internal features could include remains of aisled houses (S4, 52–4). The Roman villa at East Wear Bay, Folkestone, first excavated in 1924, is now threatened by cliff erosion and is currently under further excavation (SE2. 91; Selkirk 2012). Coastal religious sites include the Hayling Island temple, which is associated with nearby inhumations, cremations, and pits (SE1, 15). The collection of Roman finds from Nornour on the Isles of Scilly suggests that the site was a shrine to a native marine deity; they include brooches, coins of 1st–4th century date, glassware, beads and small clay Venus-like figurines (IoS, 73 and 101). A conventional Romano-British temple at Jordan Hill, Weymouth, Dorset overlooks the sea but its attribution to any marine deity is impossible (Fig 4.16).

Anglo-Saxon

The most conspicuous Anglo-Saxon monuments on the coast are religious houses (frequently refounded, rebuilt and expanded after the Norman Conquest) and stationary fish traps, both of which are considered separately below. Evidence for agricultural, domestic and burial activity on the coast is more limited. At Green Shiel, Holy Island, five rectangular buildings with stone footings, probably to support turf walls, have been excavated (O'Sullivan and Young 1991; NE1, 213; *see also* Fig 4.17). One paired structure is almost 40 × 5m in internal

*Figure 4.16
Jordan Hill, Weymouth, Dorset. Foundations of a Romano-British coastal temple seen in snowy weather.*

Figure 4.17
Green Shiel, Holy Island,
Northumberland. Remains
of a 9th-century domestic
stone building.

dimensions. The abundance of immature cattle bones was thought to indicate that the site may have had a specialist stock-rearing function. The site produced 9th-century coins and appears to have been abandoned by the 11th century. The Anglo-Saxon inhumation cemetery at Bamburgh, known as the Bowl Hole, was first exposed by dune erosion in 1817, and excavations were also undertaken in 1998–2007. East–west aligned burials imply 6th-century Christian use, though there may have been an intrusive later pagan phase. The site remains at risk of erosion (NE2, 18). Field survey detected possible artificial burial mounds within the dune system (NE5, 229). An early Christian and Anglo-Saxon cemetery, dated by radiocarbon to the mid-7th to late 9th centuries, at Seaham is 130m from the cliff edge; the church of St Mary itself has a late 7th–8th century nave (NE2, 35). Fragmentary stone crosses, one of 10th-century date, have been reported from Alnmouth, near St Waleric's church and from the river at Warkworth (NE1, 176).

In Yorkshire, between Whitby and Reighton place-name evidence suggests a dispersed pattern of farming settlements, with the monastic foundation at Whitby and larger proto-urban centres developing only where there were natural harbours or defensible headlands, at Whitby, Scarborough and Filey (YL1, 92). The present pattern of villages in Holderness is though to perpetuate a settlement pattern established in the Middle Saxon period, but archaeological evidence is slight, apart from early to mid-Anglian inhumation cemeteries at Sewerby and Hornsea (YL2, 182–3). A cart burial probably of Anglo-Saxon date came from near the seafront at Hornsea, close to the known Anglian cemetery (YL2, 76). There is very little archaeological evidence for Anglo-Saxon activity along the Lincolnshire coast, though numerous place-names date from this period, some with obvious Scandinavian derivation (including numerous '-by' and '-thorpe' place-names). A wattle hurdle from the beach at Sutton-on-Sea has been radiocarbon-dated to the late Saxon period. However, given the high rate of erosion along this coast it is likely to have been related to inland activity of some type, rather than being a fish trap (YL7, 49).

The aerial photographic evidence for coastal early to middle Saxon presence is generally slight in East Anglia, though at East Ruston in Norfolk thirty rectangular features, possibly grubenhauser, associated with other cropmarks have been identified (N2, 115–16). Cemeteries are known from the Roman fort sites at Burgh Castle and Caister-on-Sea, the former extending into the middle Saxon period and perhaps associated with a church or monastery. A late 5th- to early 6th-century gold bracteate brooch was recovered during excavation at Blakeney Freshes. It could have been produced in Scandinavia or even Kent. The find had no context, and may simply represent a casual loss, though in Kent bracteates are generally grave goods (Birks 2003).

At Covehithe, Suffolk, a possible Anglo-Saxon settlement and cemetery are visible as cropmarks in the vicinity of St Andrew's church, about 400m from the present cliff edge: loss of the site to erosion in about 70 years is predicted. There may have been relocation of settlement to the east in the mid–late Saxon period, as in other locations (S4, 66–70). Late Saxon sites at Gisleham and Covehithe are suspected from artefact scatters close to churches (S3, 38). An excavation by Basil Brown detected a long building with probable grubenhauser and Anglo-Saxon pottery in Butley, at 15m OD overlooking Butley Creek, with further settlement evidence to the north. Early cemetery sites occur at Friston and Snape – the latter with three boat burials and cremations, some with hanging bowls (S3, 54). At Sutton Hoo, Suffolk, the very high status, probably royal, cemetery of the late 6th to early 7th centuries AD overlooks the Deben Estuary. Carver (1998) interprets the lavish burials there as, in part, a reaction to the arrival of the mission of St Augustine in 597, and the conversion of the kingdom of Kent to Christianity, with the implied extension of continental influence into England. The burials, especially, the great ship burial, were emphatically pagan, and were assertions of power and authority. The campaign of excavations in the 1980s also detected groups of satellite burials around some mounds, which are interpreted as those of execution victims and could be seen as those of individuals who in some way were 'ideological or political deviants', executed at a *cwealstow*, or 'killing place.'

French pottery in early to middle Saxon Kent was later replaced by Ipswich ware, with a distribution suggesting coastal trade (SE3, 50). Along the New Forest coastline Anglo-Saxon settlement is indicated by an enclosure including a grubenhaus and other features at Milton Manor (NF1, appendix A). St Mary's, Wareham, has produced 6th–9th century gravestones with early Christian inscriptions,

cut into reused Roman stone masonry. The church later became a Minster church: Wareham became a burgh in the 9th century. The surviving ramparts protected the town on three sides, the fourth being the River Frome. They were later reinforced with a stone wall on top, perhaps in the late 10th–early 11th centuries (Dyer and Darvill 2010, 86–8). Around the Rivers Fowey and Fal, early medieval religious sites include enclosures known as lanns, often in creekside locations and sited to maximise use of the river for navigation, for example at St Winnow, located next to a deep channel (Parkes 2000, 10–11; Ratcliffe 1997, 18). In the North-West continued activity has been recorded at former Roman forts at Lancaster, Carlisle, Ravenglass and Muncaster but other evidence for settlement is sparse (NW1, 54).

Distinctively Viking-style burials are uncommon although, in the 19th century, inhumations of a woman and two men with a bone comb and a 10th-century enamel brooch were excavated from an unlocated barrow in the North-West, either at Bedlington or Cambois (NE1, 176).

Medieval

Coastal religious foundations, castles, salterns, land-claim and ports are discussed thematically below, so this section will focus on the evidence for rural settlement and agriculture. Aerial survey in the North-East recorded extensive ridge-and-furrow, especially in Northumberland, most of which has now been levelled by modern farming or built over (Fig 4.18). Evidence for earthwork features of settlements was sparse, confined to a deserted medieval settlement at Budle, a monastic grange at Fenham and manorial complexes at Easington and Saltburn Maske (NE 4, 22–3). Ridge-and-furrow is also widespread in Yorkshire, and the deserted settlements at Reighton and Speeton still survive as visible earthworks (YL1, 85). Some deserted settlements are actively being eroded, as at Rolston and Cowden (YL6, 99). Further south numerous modern villages, such as Withernsea, are located inland from their medieval predecessors, owing to losses from coastal erosion: here the present church, completed 1488, replaced the original lost in the 15th century (YL2, 183–6). Records dating to the 16th century indicate the loss of 38 houses at Hornsea between 1547 and 1609,

Figure 4.18
Berwick-on-Tweed,
Northumberland. Ridge-
and-furrow preserved
within the golf course.

with erosion continuing into the 19th century and beyond (YL2, 78–9). The medieval settlement of Owthorne, including its church, manor house and parsonage, was lost by 1844 (YL2, 111).

Between Donna Nook and Gibraltar Point, extensive areas of ridge-and-furrow have been mapped from aerial photographs and settlement earthworks are known from a number of locations. The principle medieval settlements lie around 1km inland from the coast, and there are relatively few access roads across extensive salt-marsh to the beach. These were presumably related to the use of beach-launched vessels and salterns (YL3, 35). At Theddlethorpe the earthworks of a deserted settlement appear to have been associated with a small inlet or haven, now disused, but cloth seals, recorded ship losses from the vicinity, two cannon at The Hall, and a 19th-century signal staff imply that it may have functioned until comparatively recent times (YL3, 36). An association of medieval or post-medieval earthworks with 90° bends in the medieval Sea Bank at Anderby, and possible settlement earthworks near a small inlet at Huttoft also imply former small havens (YL3, 44–6).

In Norfolk, the village of Eccles-next-the-Sea was gradually destroyed by coastal erosion from the 17th century onwards, owing to dune recession. St Mary's church tower survived until 1895, when it collapsed. Following storms, the rubble from the church is occasionally exposed, along with skeletal remains and

other features associated with the village (N2, 11). Other churches, including St Mary's Happisburgh, are at risk of future erosion (Fig 4.19). St Michael's, Sidestrand, was largely demolished in 1880, except the tower, which collapsed in 1916; the last section of churchyard wall fell over the cliff in 1936 (N1, 28). Other lost villages include Little Waxham and Shipden (N2, 126). Earthworks and cropmarks of some 30 sites were recorded by the National Mapping

Figure 4.19
St Mary's, Happisburgh,
Norfolk. The current SMP
policy for this location is
'No Active Intervention', so
the church is at risk from
erosion in the longer term.

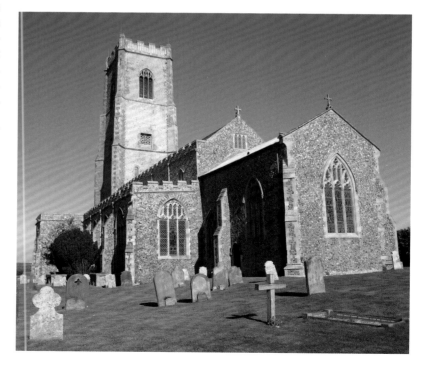

Programme in the coastal zone, including enclosures, building platforms and hollow ways, besides enclosures related to crofts and tofts. Little Ringstead, near Hunstanton, may have been abandoned in the immediate post-Black Death period. The church sits within a system of rectilinear ditches, defining tofts, crofts and fields (N2, 125–30). In total, 45 'moats' were also identified, including a well-defined example at Roughton consisting of a four-sided trapezoidal enclosure, with a central platform 90 × 68m and internal features (N2, 130–3). Almost 100 locations of ridge-and-furrow were recorded, especially in the west of the county, perhaps largely related to land drainage, for many sites are in low-lying locations (N2, 133–8). V-shaped ditches seen in a cliff face at Hunstanton were thought to be part of the field system recorded by the National Mapping Programme (N1, 58). Other landscape features include abandoned roads, including those in later landscape parks, at Holkham. In Suffolk, fieldwalking at Easton Bavents, Covehithe, and other locations produced concentrations of medieval artefacts, most probably indicating settlement areas (S1, 35–7). Easton Bavents parish has been largely lost by cliff erosion: medieval material has been recorded from the cliff edge (S3, 39).

Medieval and post-medieval agricultural features in Kent included field boundaries (in the Stour Valley), ridge-and-furrow, enclosures, a Deserted Medieval Village at Spruckleham and numerous hay-stack stances, notably in the Stour, at Sandwich and in the Lydden Valley (SE1, 46). The majority of medieval sites elsewhere in the South-East were related to ridge-and-furrow – as on Portsea, Gosport, North Binness Island, Brockhampton, Stoke Common and between Portchester Castle and Fareham (SE1, 17). Many probably post-medieval fields of ridge-and-furrow were defined around Havant and on Hayling Island, Fareham and Farlington Marshes (SE1, 20) and at St Catherine's Point and Newtown (IoW, 63–5 and 99). In the North-West cultivation terraces and lynchets were noted, especially around the mouth of the River Kent in Morecambe Bay. Ridge-and-furrow was widespread, notably in South Ribble and Barrow in Furness (NW3, 21–2). At Arnside remains of a longhouse, enclosure and associated lynchets mark a deserted settlement (NW1, 135).

Early Modern and Modern

Since the amount of information available for these periods is so very much greater than for earlier ones, they are best considered thematically discussing ports, quays and ship-building, fishing structures, natural resources including salt, wildfowl and seaweed, extractive industries, military defence, control, navigation and rescue, religious foundations, tourism and hulks and wrecks. The non-urban and non-residential coastal historic environment of the 20th century is overwhelmingly dominated by military structures and sites. Furthermore, upwards of 80 per cent of sites recorded during the National Mapping Programme (NMP) surveys for the RCZAS were also military in most regions.

Coast-specific sites: ports, quays and shipbuilding

The RCZAS were never intended to attempt substantial reconsideration of the development of the major ports. They have received detailed attention from architectural historians and other researchers. However, the expansion of the Liverpool Docks from 1700 onwards is illustrated in NW1, fig 5.9. For a short review of the archaeological and historical evidence for port development, *see* Murphy (2009, 84–103). In general the Shoreline Management Plan option for urbanised coastlines is 'Hold the Line' (*see* Chapter 6) and so change is more likely to be initiated by redevelopment than coastal management. For present purposes a more limited study area for ports and other developed coasts was instituted, including the present intertidal zone seawards of defences and an area behind the existing sea-defences, perhaps 100m deep, that might be impacted by sea-defence changes in the short to medium term. Rural havens were, however, commonly recorded. New data from the RCZAS therefore relate primarily to quayside structures at the major ports and to more isolated small landings, recorded from aerial photography and ground inspection. There is only space to refer to a few examples here.

The medieval town defences of Berwick-upon-Tweed covered an area some 50 per cent greater than the better-known 16th-century artillery fortifications. It was originally the main port of Scotland, founded by King David I in the middle of the 12th century. The medieval wooden bridge crossing the Tweed is known to have been about 75m upstream from the present bridge and some parts survive (NE1, 221). Much of the coast between Whitby and Reighton consists of high cliffs but ports of varying sizes developed at locations where there were sheltered natural havens. Excavations at Whitby show that the refounding of the abbey in 1078 was associated with development of a planned town and harbour on the River Esk. Field inspection suggested that parts of the medieval precursor of Burgess Pier survive. On the West Pier a system of capstans and rope-pulleys and relays for 'warping' vessels into harbour and manoeuvring them once inside was recorded, apparently originating in the 18th century. The partly restored capstans still extant are deteriorating (YL5, 87). Survey of the harbour of Bridlington has defined its development since the 18th century and has made recommendations for conservation of its early features, including a 19th-century ship's capstan, bollards and the compass rose at the base of a weathervane (Brigham and Fraser 2012b).

Minor havens, staithes and jetties are very poorly recorded archaeologically, though there are thousands of them. Many small landings could well have been in use since at least the Bronze Age. They have been almost ignored by archaeologists, although the coastal trade in agricultural and industrial products has always been considerable and, cumulatively, the goods transported to and from them were of enormous economic significance. Offshore anchorages and beach landings may be marked by little more than concentrations of distinctive artefacts or exotic ballast stone. For example at Quarr Beach and Ryde Middle Bank off the Isle of Wight, late and post-medieval pottery assemblages included a high proportion of Breton wares, which are rare at inland sites. The inference is that these ceramics were not traded, but rather were discards from vessels. Ballast stones at Wootton Haven had sources in Cornwall and South Devon, the latter being an area where the medieval Quarr Abbey had economic interests (Tomalin *et al* 2012, 380 and 430–3).

A small stone-built quay at Dunstanburgh Castle, 72m long, has been surveyed: 14th-century sources establish that vessels were berthed at Dunstanburgh (NE1, 184). Dating is problematic, but it is thought to be a small medieval harbour (NE1, 69). Minor

havens on rocky coasts in the North-East were in general natural inlets, exploited by fishing communities, although some, such as Nacker Hole, were provided with small quays for lime exporting in the 18th century, as was the harbour on Holy Island (NE1, 229). In addition the alum industry of the North-East and North Yorkshire had associated docks, essentially cuts into the local bedrock. At Osgodby at the northern end of Cayton Bay 'Johnny Flinton's Harbour' is marked by an area of foreshore cleared of stones with a small breakwater or mole, probably related to fishing (YL5, 63). At Flamborough a harbour of two piers, made largely of glacial erratics, was first mentioned in a document of 1400–01, but probably is of 14th-century origin. It fell into disuse in the 16th century and thereafter was largely robbed for construction elsewhere. Parts of the facings (c 10m apart) survive and have been planned in detail (Brigham and Fraser 2012a).

On an estuarine coast, the Eau Brink Cut was constructed in Norfolk in 1821 to straighten a bend in the river towards Kings Lynn. It was extended in 1853, with associated embankments for reclamation (N1, 12). Ground survey recorded post alignments, 800m and 1.5km long, thought to be related to this cut (N1, 53).

Between Donna Nook and Gibraltar Point in Lincolnshire there were minor havens at North Somercotes, Grainthorpe, Theddlethorpe St Helen and Huttoft. Natural inlets appear to have been improved by means of protective banks, and modified to provide berthing. From Theddlethorpe there is a record of a find of two lead cloth seals, probably related to a 17th-century wool trade (YL3, 69). Wainfleet Haven, at the mouth of the River Steeping, was an important port locally from the medieval period onwards, though movement of sand bars around Gibraltar Point resulted in its demise by the 1920s (YL4, 24). The beach-launched fishing industry and trade via these small havens declined in the 19th century, activity transferring to larger ports and towns such as Grimsby and Skegness (YL7, 50). Beach landings, where there were gaps in the dunes, were locally known as 'pullovers' (YL7, 16).

Numerous features related to harbours and quays were recorded in Norfolk, including a brick harbour wall at Brancaster Staithe, a flint wall at Salthouse, and a brick and flint storehouse at Thornham. Timber revetments, jetties and quays were recorded along the River

Great Ouse, in the North Norfolk marshes, and in Breydon Water, with hards at various locations in North Norfolk. It is thought that most are post-medieval in date (N1, 150). In Suffolk, surface shoreline scatters of Roman artefacts at Walberswick and a cremation cemetery at Orford imply the former existence of Roman ports, and whole pots from the Alde, at Iken, suggest a wreck or quayside (S4, 11). Of the 484 records made in the Suffolk estuaries, some 25 per cent appeared on 1st–3rd edition OS maps, but some 19th century and later structures provide access to modern farms that had medieval precursors and which themselves would have needed river access. In some cases, for example at Methersgate on the River Deben, medieval ceramics and peg-tiles were associated with modern features (S1, 46). Hards in these estuaries are simply spreads of material emplaced for beaching, comprising oyster shell or rubble, sometimes including coprolite (S1, 19), though in some locations small docks were dug into the salt-marsh, as at Boyton on the Ore, constructed around 1780 (S1, 23). At Blythburgh medieval material and timber jetties (possibly more recent) mark a probable harbour area. Buss Creek, Southwold, may also have been a harbour area, having produced ship remains and other timbers (S3, 43).

Post-medieval port development depended on approaches by river to their quays. Later quays and hards have been identified from map and aerial photographic evidence at numerous locations, including Redgate Hard, Wherstead and Waterhouse Quay, Arwarton. At Sudbourne an earthwork causeway and related structures indicate a ferry crossing (S4, 115–16), confirming one shown in a map of 1601.

Landing points around Foulness in Essex around the north of Canvey Island also comprised lines of large timber posts: they served individual farms (E1, 15–16; E5, 3–6). At an earlier date, at Stanford Wharf, an early Roman structure represented by twelve oak piles, some 13m long, adjacent to a tidal channel and with a tapering end, is interpreted as a boat-house. It could have housed a small estuary barge, similar to that from Barlands Farm, perhaps used in conjunction with the salt industry at the site (Biddulph *et al* 2012, 101–2). In the Medway there are numerous timber wharves and jetties, often associated with industrial activities. Chalk hards, with timber retaining structures, for unloading barges at low tide also occur (NK5, 23 and appendix 2). There are numerous post-medieval wharves, piers, quays and jetties in the Gravesend area, in part related to the town's functions as the base for ferries, a provisioning point, and customs control (from 1356). Landing stages and steps, piers, jetties and wharves are recorded but most, apart from the main pier itself are no longer operational (NK6, 12; NK6, 73–4).

On Portsea Island in Hampshire boat hards, jetties and a wharf were mapped from aerial photographs, besides the sea lock and basin for the Chichester Canal (SE1, 20–1). There is a scatter of artefacts of Roman, medieval and later date from anchorages in the Solent, discarded in antiquity and now recovered from oyster dredges at Ryde Middle Bank, Spithead and Mother Bank (Loader *et al* 1997, 5 and 18). At West Cowes structures include shipyards, slipways, a stone-faced quay, the Medina Ropeworks and a malthouse – besides the Royal Yacht Squadron (IoW, 112). At Gin's Farm on the Beaulieu River there was a medieval harbour (Fig 4.20), presumably constructed by the Cistercians of St Leonard's Grange (NF1, appendix C). 'Gin' is supposed to have derived from *inginium*, a mechanical device, most probably a crane. The Dorset coast also shows numerous minor causeways, small harbours and landings, many short-lived. In Poole Harbour two structures in the south-west part of the harbour, 100m and 55m long supported on piles and capped with Purbeck stone slabs, are interpreted as causeways involved in the export of iron, shale and ceramics produced on Furzey and Green Islands. The features are dated to around 250 BC (Dyer and Darvill 2010, 39–40).

At Castle Dore on the Fowey, Iron Age contexts have produced glass bracelets, lump galls and amphorae, probably from landings close to the St Catherine's Point cliff castle (Parkes 2000, 25). Lostwithiel on the Fowey Estuary was the base for shipping and taxing of tin exports at the complex of buildings known as the Duchy Palace by the dukes of Cornwall. In 1203–1205 it was 7–8th of 35 ports in England, but later declined as tin production decreased and the port was silted with sediment derived from tin mining (Parkes 2000, 13). In the later Middle Ages Fowey became the main port. The main harbour on Scilly in the early Middle Ages was probably at St Helen's Pool, where the monks of Tresco levied tolls, with a

Figure 4.20
Gin's Farm, Hampshire.
This was the site of a
monastic quay. Some
features of the harbour
survive as earthworks
in the foreground.

medieval quay at Old Town. By the 16th century the settlement focus had shifted to The Hugh, where the quay was built in 1601, extended for military purposes in the 1740s, with a new deep-water commercial quay in the late 1830s. Numerous other small quays, slipways and piers occur around the islands (IoS, 117–89).

Grange Pill on the Severn is thought to be an artificial cut of the 12th–13th century. A Lower Quay, of stone and timber, 12 × 5m has a dendrochronological date of 1172. The Upper Quay is larger and dated by dendrochronology to after 1100, with a possible repair in 1206. Construction was probably by Tintern Abbey (S1, 24). At Meols on the Wirral 3rd-century Carthaginian coins suggest long-distance trade, as do mainly 1st–2nd century AD coins and other artefacts (NW1, 73) and a probable early medieval trading port was exposed by sand dune recession in the 19th century. Artefacts of 7th century and later date indicate contacts with the Mediterranean and Africa. The site may have been a Scandinavian trading port in its later phases. Skinburness in Cumbria was founded in the 12th century to supply the English fleet of Edward I invading Scotland, but had been largely eroded by 1305 (Fig 4.21). Earthworks and structural features probably relating to it still survive (NW1, 194).

Shipbuilding has been undertaken at many locations, but from the 19th century onwards the industry was focused in the north, with ready access to the raw materials of coal, iron

and steel (Murphy 2009, 66–70). The extant evidence for iron and steel shipbuilding in the industrial areas of the north – including the Tees, Wear, Tyne, Mersey, Whitehaven and Barrow in Furness – comprises principally the 19th–20th century shipyards, now often redeveloped. Field survey also recorded the remains of infilled dry-docks and slipways along the River Esk related to Whitby's significance as a boat-building and repairing port. Some may have originated in the 18th century, but most were infilled in the early 20th century (YL5, 34). It is probable that these infilled docks will be well preserved structurally: excavation has demonstrated survival of ship timbers in one at Church Street (YL5, 37).

Timber shipbuilding in the south, for example at Fishbourne on the Isle of Wight (IoW, 118) and at Buckler's Hard (NF1, 28; NF4, 42; Fig 4.22), frequently has a less obvious character. Timber piles for shoring and staging have been recorded at Fishbourne, dated to after 1470 AD (Tomalin *et al* 2012, 296–300). Sites include a saw-pit at Warsash and timber ponds at New Slipper Mill, Emsworth (SE2, 55). Physical remains of timber shipbuilding are uncommon, though in 1986–7 ship timbers – new and unused – were recorded on what was originally the shore, near St James Church, Poole. They comprised cut knees and V-shaped timbers, cut for clinker planking, dating to the late 14th–15th centuries (Dyer and Darvill 2010, 146–63).

Figure 4.21
Skinburness, Cumbria.
Medieval features including
house platforms relate
to a port from which
the invasion of Scotland
was supported in the
13th century.

Figure 4.22
Buckler's Hard, Hampshire.
Substantial vessels,
including HMS
Agamemnon, *were*
constructed here using
New Forest oak, from 1698
to 1827. The slips still
survive as earthworks.

Coastal-specific sites: fishing structures

Stationary fish traps are widespread. The principle is simple, and persisted for millennia: a timber or stone obstruction, built in an estuary or sheltered shore, most often V-shaped and with its wide mouth facing upstream, will catch fish on a falling tide. Palaeochannel sediments at Hartlepool Bay have produced a hurdle panel from the early 4th millennium cal BC, tentatively interpreted as part of a fish trap (Waughman 2005); at Must Farm, Cambridgeshire, Bronze Age V-shaped weirs and baskets, probably for catching eels, have been recorded; and there is a conical trap of early Bronze Age date from Holland in the Humber wetlands. But in general traps were small and infrequent in England compared to later periods (Bell 2013). The paucity of evidence for prehistoric stationary fisheries may perhaps have been related to the use of the rivers and sea for burial, which would have imposed a cultural prohibition. The absence of Romano-British fish traps is unexplained.

Examples of fish traps come from many parts of the country (Fig 4.23). A stone-built rectangular structure, some 30m across, at Nova Scotia inlet, Dunstanburgh, has been interpreted as part of a fish trap (NE1, 185),

and stone structures, thought to be fish weirs, have also been recorded in Budle Bay (NE1, 228). In addition a large stone fish trap at Whitburn, south of Newcastle, was recorded; it resembles stone structures from Budle Bay and may be of medieval date, associated with a monastic grange of Lindisfarne Abbey (NE5, 270). Fish weirs on the foreshore at Cleethorpes are of unknown date. They include a linear stone bank, arcuate at its northern end. Further south there are V-shaped wooden structures, presumably fish weirs, some 50–100m long (YL2, 156). At Holme Beach, Norfolk, a complex of Anglo-Saxon timber fish traps has been recorded on an open coast (Robertson and Ames 2010). Elsewhere, submerged V- and L-shaped linear features within Blakeney Harbour are most likely to be fish traps (N2, 220–5).

Fish traps in Suffolk include the massive V-shaped feature at Holbrook Bay (Fig 4.24), almost 310m in length (S4, 61–3). The structure comprised multiple lines of parallel posts, with horizontal wattling in places alongside them (S2, 4–7). After Bayesian processing, samples gave radiocarbon dates of cal AD 680–850 (at 95% probability), or cal AD 630–690 (at 68% probability), with later dates of cal AD 880–1025, probably indicating a later repair. Wooden structures, including wattle panels,

Figure 4.23
St Bees, Cumbria. A stone-built fish trap associated with an adjacent monastic community.

Figure 4.24
Holbrook Bay, Suffolk.
Part of a massive timber
fish trap involving multiple
lines of stakes.

dating 420–590 cal AD, were recorded on the Deben at Sutton Hoo, and are interpreted as fish traps (S3, 62). Further south in Essex, notably at Collin's Creek, The Nass and Sales Point, timber fish traps have provided radiocarbon dates focusing between 600–900 AD (Murphy 2010, 218–19). A probable medieval/post-medieval fishtrap recorded at Peg Fleet, Kent, comprised 43 vertical stakes, over some 20m, with woven brushwood (NK5, 28), and there are further probable structures on Nor Marsh (NK5, 37). Another at Shornemead comprised vertical stakes with horizontally laid wattle panels and a possible fish basket (NK6, 12). Others at Damhead Creek comprised six rows of stakes (NK6, 15–16).

Probable fish traps around Chichester Harbour and Thorney Island were generally linear lines of posts at 90° to the shoreline, but one V-shaped trap was defined within the Thorney Channel (SE2, 61–2). On the Isle of Wight a V-shaped fish weir was built at Binstead, dated to cal AD 890–1040 and cal AD 810–1020. Largely of timber, though supported by stone at the base, it had leader arms at least 128m long with an unusual circle of stakes with associated hurling (Tomalin et al 2012, 217–21). At Quarr there were further traps in the 14th century and there was a stone and timber V-shaped fish weir at Ryde (IoW, 19). Off Need's Ore Point a group of small V-shaped fish weirs,

about 10m long were mapped from aerial photographs (NF3, 35). A medieval or post-medieval fish trap, comprising a U-shaped line of stones, was mapped just south of Portchester Castle (SE1, 16). However, fish traps were uncommon in this area (SE1, 13–14).

Phase 2 of the Severn RCZAS (S4) was very largely focused on stationary fishing structures (Chadwick and Catchpole 2010). Many V- and U-shaped structures identified in the NMP survey were interpreted as traps, but during fieldwork some were redefined as net and line fishing weights for upright net-hangs. Related structures in Somerset were stone rings or doughnuts that supported vertical uprights. On cobbled beaches net-hangs are identified from narrow lines of stone clearance. These may be difficult to distinguish from ground line gullies, where lines of baited hooks were set out perpendicular to the shore. Most net-hang lines are thought to be of 19th and 20th century date, but in spite of their recent date they may perpetuate lines in use for generations. Herring, mullet, skate, whiting, sole and sprats were caught in Minehead Bay; other species further upstream, with salmon and shad in the upper estuary.

Stone-built fish weirs occurred mainly in Somerset. Generally V- or U-shaped with the open arms facing landwards, they mainly caught fish on ebb tides. They are variable in size, some with leader arms 10m wide and hundreds of metres long, down to smaller ones, 20–30m across. At the apex is a channel or gut, which may have everted 'horns' at the end, where catch baskets or bag-like nets were fixed by posts. Some were carefully built as drystone walling, but others were merely rubble. Some exploited natural features and some weirs were linked W-shaped structures. Weirs higher up the shore functioned only on spring tides. Tidal erosion is the main cause of degradation, though some were buried beneath sediments, and later robbing for new construction also occurred. The surviving examples are probably post-medieval or early modern. Other stone-built features included 'ponds', conger eel traps and lines of cairns.

Wooden fish traps were recorded at Beachley and Aust/Oldbury. Typically V-shaped post lines, some still include wattle panels and fishing baskets, fixed by withy ties. Fish were often caught on waxing incoming tides, but others faced upstream to catch fish on an ebbing tide. At their apices there were circular catch baskets in some cases. At Woolaston, radio-carbon dates on fish baskets and a stake and hurdle structure indicated use between the 8th and 11th centuries, presumably relating to construction and repair. At Aust/Oldbury Flats there was a T-shaped structure of post-medieval date, whilst other V-shaped structures were of late 7th–early 8th century. Other V-shaped structures suggest construction and use from the 11th–13th centuries. Small stake-built structures housed fish baskets known as putcheons, used to catch eels, lampreys and lamperns. Other basketry from these sites may have consisted of frails, used to transport fish away. On Stert Flat, V-shaped or tick-shaped structures caught fish on an ebbing tide, some dating to the 8th–13th centuries.

Putts (basket traps placed in weirs in single tier ranks) are probably mainly medieval. Putcher ranks, with baskets in long ranks of 3–4 tiers high) were introduced in the 18th and 19th centuries, and were used to take salmon. One of the latter is still in use at Awre. Degraded remains in general show as two lines of posts at right-angles to the shoreline. On rock outcrops postholes were cut. Owners included the Crown, gentry and monasteries, as indicated by sales of rights in 16th-century and later documents. The fishery was regulated by the Salmon Fisheries Acts 1861 and 1865, to prevent unlicensed structures depleting stocks and being a hazard to navigation.

In the North-West, stone and wooden fish traps were often associated with monastic/ecclesiastical sites (NW3, 39 and fig 38). V- and W-shaped structures focus around Morecambe Bay and extend northwards into Cumbria. The trap at Plover Scar lighthouse and those near Cockerham may have been associated with Cockersand Abbey (NW1, 136). Stone and timber traps at Cowp Scar, Cartmel Peninsula, are dated to 575±28 BP or 1303–1368 cal AD (NW1, 136). Fish traps of presumed medieval date at St Bees may have been associated with the Priory (NW1, 163). In the Solway Firth a distinctive form of stationary fishery is haaf-netting, reputed to date back to Anglo-Scandinavian times, which plainly would leave no archaeological trace on the shore (Fig 4.25).

Figure 4.25
Solway Firth, Cumbria.
A haaf-netter.

The probability ranges of radiocarbon dates on wooden traps in the East of England and the Severn imply activity from the 7th–13th century, and again in the post-medieval period, with a concentration of dates in the 7th–9th centuries (Chadwick and Catchpole 2010; Murphy 2010). Fish traps are often associated with nearby monastic foundations and so could be considered the precursors of monastic fish ponds. Locally, their subsequent changes in importance could have been related to fish-stock depletion, conflict (most notably in the 9th century and later conflicts) or to the expansion of off-shore fisheries from around the 10th century.

Facilities for shell-fisheries are likewise widespread. At Robin Hood's Bay, Yorkshire, there was a foreshore landing for fishing boats, marked by posts, associated with apparent net-drying racks and a series of hullies: square rock-cut pits that originally housed wooden boxes used as shellfish holding tanks (YL1, 42–3). 'Oyster Ponds' are indicated on a map of 1865 in the Aln Estuary and have been archaeologically recorded as lines of stakes on the foreshore, formerly revetting wooden boards to form tanks, 6 × 9.5m and 4 × 9.5m (NE1, 185). They seem to relate to a short-lived fishery, c 1794–1860 (NE5, 172). Oyster pits have been recorded at Huttoft, Chapel St Leonard, and Ingoldmells in Lincolnshire, and there is a record of probable oyster beds at Skegness.

Shellfish beds were recorded at 33 locations in Norfolk. At Brancaster Staithe, for example, a group of 11 pits, mostly timber-lined but two brick-lined, were recorded, with related sluices and revetments, reputedly dating from the 1880s (N1, 67). At Burnham Overy Cockle Strand a group of interlinked beds, irregular in form, and 1–10m across, lies just outside a seabed defence of 1822. Further inland a much more regularly cut system of beds matches beds shown on the 1825 Burnham Overy Enclosure Map, and are labelled 'mussel beds' in the OS 1927 map. Influx of sand finished the oyster and mussel industry in the late 19th to early 20th centuries. On the west coast at Heacham there are two groups of oyster pits. One comprised at least 67 beds or pits, 8–21m long and 3.5–7.5m wide. They are separated from the shore by a bank described as 'new' in 1781 (N2, 220–5). Oyster beds in Suffolk are roughly rectangular or square, but vary widely in size – from 2 to over 70m, and the lay-out varies from

ordered to 'chaotic'. Most are focused to the north of the Rivers Alde and Ore and the Butley River. Dating is problematic, though relationships with dated sea walls suggest dates in some cases (S4, 92–100).

The West Mersea, Essex, oyster grounds originated at least as early as 1046, but the main development was post-medieval, with 130 vessels by 1811 and there were 350–400 in the Blackwater and Colne by 1846. The pits associated still survive, with related structures comprising landings, walkways, layings and sluice gates (E1, 7). In the area of Foulness there have been oyster layings at The Middleway, Shelford Creek and New England Creek since the 17th century. On New England Island, pits were associated with a sluice, walkways, a platform possibly for a packing shed and a probable hoist; elsewhere on the island a 'hoove', an early form of sluice, was identified. Isolated posts may have acted as 'metes' defining the oyster layings (E1, 16–17; E5, 4). Phase 1 survey in North Kent indicated oyster pits in tributaries of The Swale (KY1, 47).

On the south coast there was oyster production at Emsworth, Langstone and Chichester Harbours (SE1, 13–14). Oyster beds of the New Milton Fishery in Langstone Harbour relate to a late phase of production in the 19th century. Oyster beds are well defined in Chichester Harbour to the north of West Itchenor and were recorded in 1946 aerial photography. There are extensive oyster beds around Thorney Island, comprising rectilinear features to the south of Prinsted. Fifty individual beds have been defined at Emsworth (Fig 4.26), related to a major focus of production in the 19th century (SE2, 54). Post-medieval salterns in the Newtown Estuary in the Isle of Wight were later reused as oyster-beds: surviving remains include post alignments, earthworks and fired clay (IoW, 99). In the Fal oysters were dredged, then stored in rectangular stone-walled structures at the mouth of the Pill and Channels Creeks, known as 'keeps'. The mussel processing plant at Lytham comprised a large concrete tank with a seawater inlet pipe in which potentially contaminated mussels were placed to clean themselves (NW1, 111).

Fish catches, of course, had to be processed for preservation and onward trade; and facilities had to be provided for the fishermen. Structural evidence associated with the Tweed salmon fishery consists of an 18th-century shiel

Figure 4.26
Emsworth, Hampshire.
19th- to early 20th-century
oyster ponds.

(Grade II listed) at Spittall (NE2, 15). Square and rectangular rock-cut features at Low Hauxley, Druridge Bay, are interpreted as bratt holes or hullies to store bait or a catch (NE5, 159–60). A site at Leigh Beck, Canvey Island, originated as a Roman Red Hill with localised settlement. It was reoccupied in the 12th–13th centuries, when deposits rich in fish remains were deposited, including shark, ray, herring, conger eel, cod, haddock, whiting, horse mackerel, grey mullet and flatfish (Wilkinson and Murphy 1995, 183–95). It evidently represents a fishing site and possibly a fish-processing site, supplying fish inland. At Stanford Wharf, also in Essex, fish processing was directly associated with salt production during the Roman period. A later Roman ditch produced an exceptional density of small fish-bones, mainly of juvenile herring, sprat and smelt: around 1000 bones per gram of sediment. The majority of fish represented were 30–50mm in length. They are interpreted as a residue from production of a salted fish sauce, *garum, liquamen* or the derivative *allec* (Biddulph *et al* 2012, 119–20). At Old Town Bay, Scilly, a rectangular trough, possibly used for salting, is recorded, while a drystone sub-rectangular building at North Hill, Samson, may be a fish-smoking house. A 19th-century shellfish bed is

also recorded, from Appletree Point (IoS, 116).

Fishing-related structures in the Severn include fishing houses, generally single storied structures with fireplaces, chimneys and a storage loft for gear (Chadwick and Catchpole 2010). Areas of stone clearance imply beach moorings. Structures directly related to off-shore fish catching are uncommon, though 'huers' huts' in prominent locations in Cornwall were the place from which watchers could observe and report the arrival of the pilchard shoals and inform the fishing fleet. In the Fowey Estuary the near-shore pilchard fishery was supported by fish cellars, where the catch was pressed, then pickled or smoked for export (Parkes 2000, 16): at Godrevy, Cornwall, a 17th-century example was exposed in a cliff section (Tangye 1991). At Berwick a whaling-oil processing shed now known as Pier Maltings (Grade II Listed) was built around 1807 and there is documentary evidence for the industry in the period 1807–1838, though this activity was in fact focused on the south bank of the Tweed at Tweedmouth. Whale bones have been recovered from the site (NE2, 15). Elsewhere whaling fleets were based at Kingston-upon-Hull, in the Thames Estuary and elsewhere, but the surviving visible remains of this 'fishery' are slight (Fig 4.27).

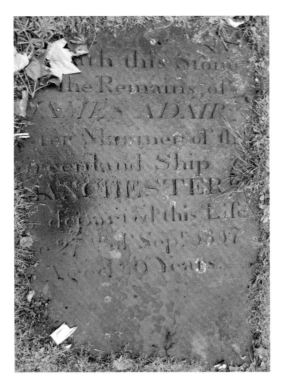

Figure 4.27
Kingston-upon Hull, Air Street cemetery. Memorial to James Adair (1837), mariner of the Greenland whaling ship, Manchester.

Coast specific sites: natural resources including salt, wildfowl and seaweed

Salt was in the past an essential commodity, used for the curing and preservation of foods, principally meat, fish and dairy products. A supply of salt helped to ensure adequate food-stores through the winter months, and could be used to convert a fish catch into a processed product suitable for trading over long distances. The raw material was essentially free, except for the effort involved in saltern construction. Salt production from seawater in England was, however, a highly energy-intensive industry and was initially confined to the summer months when lower rainfall made preliminary wind-evaporation practical. Centres of salt production have shifted through time, partly depending on fuel availability, besides the changing locations of the main urban markets. The technology of production likewise changed through time involving, initially direct evaporation by heat, and subsequently the use of the technique known as 'sleeching' in the medieval period, whereby salt water was poured through salt-rich mud to produce a strong brine. Later still, coal was used to evaporate brine produced by earlier wind-evaporation.

The earliest known saltern from Essex is that at South Woodham Ferrers, with a radiocarbon date of 1412–1130 cal BC. It comprised a hearth and a scatter of briquetage – mainly of small evaporating vessels – and a bucket-shaped pot (Wilkinson and Murphy 1995, 157–65). The middle Bronze Age settlement at Brean Down has produced the earliest briquetage from England (S1, 44; Bell 1990).

Iron Age and Romano-British salt production is known from many parts of the country. Briquetage from North Road, Berwick, may have been related to salt refining rather than brine evaporation, since this site lies near a 45m cliff top (NE1, 224). An extensive salt industry, beginning in the early Iron Age, is focused around Ingoldmells, and finds of prehistoric briquetage have been reported from Skegness (YL3, 65). The industry continued into the Romano-British period, still focused around Ingoldmells, though with salterns also at Chapel St Leonard and perhaps the Skegness area (YL3, 65–6). In the medieval period there was salt production on the Lincolnshire coast, as evidenced by place-names such as Saltfleet, and archaeological records (YL3, 47, 68), though seemingly on a smaller scale than on the Wash coastline. Beach recharge between Mablethorpe and Skegness has resulted in sites exposed in the intertidal zone being obscured by sand cover, but salterns slightly inland are sometimes impacted by ditch-cleaning or development (YL7, 49). In Lincolnshire and Norfolk many Iron Age and Roman salterns are known from field surveys and excavations along the contemporary coastline, now inland (Lane and Morris 2001, fig 2). In terms of aerial photography, the NMP mapped five sites marked by earthworks and light-coloured soil, some of which had previously produced fire-bars and early Roman pottery, as at Sandringham. Two further mounds at Snettisham and Dersingham were linked to adjacent settlement areas by trackways (N2, 100–1).

In Suffolk, Iron Age to Roman Red Hills (mounds of red-fired earth and briquetage) are more numerous in the south of the county, approximately up to the River Blyth. There are 26 known sites, far fewer than in Essex. This could be a tribal (Trinovantian/Icenian) distinction, or related to the proximity of the market at Colchester (S4, 55–9). In Essex, the main phase of salt production, marked by numerous Red Hills, was in the later Iron Age and earlier Roman periods, in the estuaries of the Stour and Colne, Mersea Island and the Blackwater, Dengie, the Crouch and Foulness and Canvey Island and the Thames (S4, 166–96; Fawn *et al* 1990). The sites typically comprise low mounds of red earth, up to 2m high (though generally degraded by ploughing in reclaimed areas) and ranging from 0.25 to several hectares in size (Fig 4.28). Briquetage is frequent, comprising fire-bars, pedestal supports, slabs, and vessel fragments with hearths and channels. Production seems to have been most intense in the early Roman period.

A 44-hectare site at Stanford-le-Hope Wharf, Essex, next to Mucking Creek, was investigated in advance of ground-level reduction, sea-wall construction and breaching, as ecological mitigation for the London Gateway Port Development (Oxford Archaeology 2009; Biddulph *et al* 2012). It represents the most extensive excavation of a series of salterns in the area and has established that production began in the middle Iron Age (c 400–100 BC), with repeated activity in the early Roman period, up to c AD 120) and then again during the 3rd century. Early–middle Roman production was indicated by tidal channels and dug ditches sending water into gullies and ponds for initial wind evaporation. The brine was then sent to rectangular clay-lined tanks before heat evaporation. Waste material, ash and sediment, was piled to make low mounds, Red Hills, which in turn provided new working

areas. A late Roman saltern, post c 270, is within a new ditched enclosure, and is smaller in scale. A large well-constructed tile hearth (contrasting with the previous short-lived early Roman hearths) was excavated. There was no briquetage from the late Roman deposits, but they produced enhanced concentrations of lead in associated sediments, suggesting that evaporation was in lead tanks. Charcoal was abundant and, unusually, evidence for use of dried salt-marsh plants as fuel was recovered. It is thought that the salt-rich ash from hearths was probably subsequently reused during salt production. A large structure was marked by a circular gulley, with four post-pads of chalk and flint rubble and a central hearth. It is unparalleled but may have been a salt store or drying shed. This unusually large-scale excavation demonstrates that the chronology of salt production may be extended and refined where a sufficient area is available for study (*see* Fig 5.4).

Roman and medieval salterns in North Kent (KY1, 63) include salt mounds at Whitstable and salterns at Ford Marsh, Hoo St Werburgh and High Halstow. Roman salterns are abundant in Queenborough/Upchurch area, including Hamgreen Saltings, at Halstow and Elmley (NK2, 11; NK3, 11–12). Mounds on salt-marsh at Cooling Marshes are probably related to salt working, as are those at South Harty (NK6, 39 and 57). Grid patterns in crops at Cooling Marshes could be related to salt working (NK6, 64). At Efford Landfill Site in

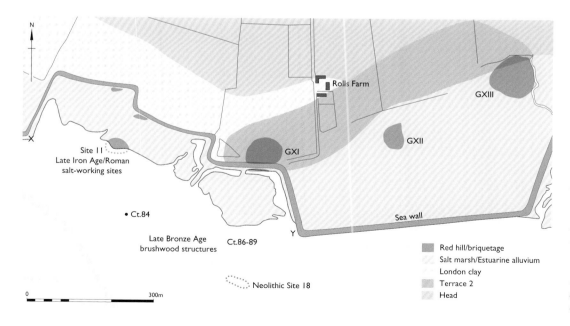

Figure 4.28
The distribution of Red Hills near Rolls Farm, Tollesbury, Essex. X–Y is a post-medieval breach of the sea wall (Wilkinson and Murphy 1995, fig 106). (Reproduced courtesy of and © Essex County Council)

the New Forest, Iron Age roundhouses and pits, with evidence for seasonal salt production, were excavated (NF1, 24). Iron Age to Romano-British salterns are reported from Fishbourne and Quarr on the Isle of Wight (Tomalin *et al* 2012, 386–7). In Poole Harbour there was widespread Iron Age to Roman salt production, partly fuelled by oil shale, and on into the early medieval period, with post-medieval salt extraction on a more elaborate industrial scale (Dyer and Danvill, 106–8). Salt production continued into the Roman period in the Severn, as at Huntspill (S1, 44).

Medieval and post-medieval salt production in the North-East is characterised by the presence of sleeching mounds, composed of waste silt. The industry expanded considerably in the 18th and 19th centuries to supply the herring fishery (NE2, 8). At Coatham East Marsh at the mouth of the Tees more than a dozen medieval saltern mounds were recorded by aerial survey over an area of around 90ha (NE4, 23). There are other sleeching mound sites around Teesmouth and the positions and extent of mounds west of North Gare were accurately mapped during field survey (NE5, 77). Documentary sources establish that there was salt production from the 12th century at Blyth, and further north at Cambois, both apparently associated with monasteries. The salt pans at Blyth became crown property in 1533 and by 1589 there were 20 pans. There were 19th-century salt works at Port Clarence, where there were also soda and chlorine works, indicating continuity with the later chemical industries of Teesmouth. By the 15th century salt was produced at Tyneside by direct boiling and this technique continued into the 18th century at Cullercoats, Blyth and Seaton Sluice. Production ceased at Blyth around 1875, but at Amble was still underway in 1887. Salt production is known to have taken place at Pan Leazes, Alnmouth, from cartographic and place-name evidence, but is poorly understood archaeologically (NE1, 103, 108, 184–5).

There is very little evidence for salt production on the Yorkshire coast between Whitby and Reighton except, perhaps, at Cloughton Wyke where the place-name 'Salt Pans' is unsupported by any archaeological evidence (YL1, 53). However, further south, in Lincolnshire, post-medieval salt production resulted in lines of saltern mounds along the contemporary shore, later perpetuated by sea-banks, notably in the Humberston/Tetney area,

now about 1km inland (YL2, 190). Along the western shore of the Wash there are saltern sites at Wrangle, with some at Old Leake and Benington. A line of saltern mounds or 'Tofts' is shown on the 1824 OS Map running from Wainfleet to Friskney and Wrangle (YL4, 23). One of the Wrangle sites produced early medieval pottery (late 11th–12th century), but fieldwalking shows the majority were of later medieval date, continuing in some instances into the 17th century (YL4, 65). Between South Wootton and North Lynn in Norfolk, there is a continuous line of saltern mounds, flanking the former course of the Great Ouse, prior to its canalisation in 1821. The industry was established by 1086 and certainly the Late Saxon sea bank cuts through salterns; else-where later sea banks link salterns. Some later structures, such as West Lynn church and St Edmund's Chapel at North Lynn, were built on saltern mounds. In the Breydon Water area, close to the medieval and later fish-curing industry at Great Yarmouth, locations of some 39 sites were defined. As in The Wash, farms appear to have been built later on saltern mounds. The evidence suggests a smaller-scale industry in the medieval period (N2, 158–63). However, by the 16th and 17th centuries, salt works were developed at Yarmouth, most notably on Cobholm Island under Nicholas Murford in 1635, where an area of some 24 hectares was devoted to allowing brine to evaporate in shallow pits in the summer months and then boiling to produce white salt (N2, 160–3).

Mounds along Essex estuaries, as at Saltcoat, Stow Maries and around the Blackwater, probably represent medieval and later salterns, though there has been no archaeological investigation (Wilkinson and Murphy 1995, 197). In East Kent 41 probably medieval salt mounds were mapped around the River Stour between Wall End and West Stourmouth, often associated with enclosures and ditches for water management (SE1, 46). At Docker Hill on the Wantsum Channel a salt mound was subsequently modified to become a moated site. Post-medieval salterns were recorded at Hamble le Rice, Newtown, Hook Park and Newton, principally by large evaporation ponds. There was a large example at Apuldram, active until 1840, and now converted to Chichester Marina (SE2, 56). The Gosport and Fareham coastal plain was not a favoured area for salt making: the only major location in the

19th century was at Fareham Creek and Hill Head (SE1, 12). Between Havant and Emsworth there are a few medieval salterns, but the main period of activity was 1600–1900. Salterns were recorded at Tipnor and other saltern sites, for example, Great Salterns, appear on the 1870 OS map (SE1, 16). An area of post-medieval salterns was mapped at Wicor Mill, Portchester (SE1, plate 4).

There are 19 known salterns in a small area to the north of Norman's Bay, Pevensey Levels (SE2, 92). Post-medieval salterns in the Newtown Estuary on the Isle of Wight were later reused as oyster-beds. Surviving remains include post alignments, earthworks and fired clay (IoW, 99). Salterns were also recorded in the Medina Estuary: features including banks and post alignments may be associated (IoW, 118).

Possible early medieval salterns are reported from Great Marsh, Exbury and Asklett in the New Forest (NF1, 26–7). Medieval examples are known from Great Marsh, Iley Dock and Hurst Spit and possibly Keyhaven, (NF1, 7–8), and between Lepe and Hurst Spit there was a focus of sites. In total, 23 were recorded from aerial photographs, being well exposed in the marshes between Lymington and Keyhaven. They are apparent from systems of drains and channels, some of a formal herring-bone form, though some comprise sinuous ditches, providing conditions suitable for sun and wind to allow evaporation prior to boiling (NF3, 34). The shift from medieval sleeching towards industrial evaporation occurred during the 17th century here, in turn resulting in large-scale land reclamation. Lymington was a noted salt-exporting port. The industry declined in the 19th century, though persisted into the early 20th, with salterns at Gin's Farm and Great Marsh. Salterns are also known from Dibden, Ashlett, Hythe and Totton (NF1, appendix E). Evaluation excavations at the Creek Cottage saltern near Lymington were focused around a salt boiling house and a storehouse, both probably of 18th-century date (Fig 4.29). Finds were mainly of post-medieval date, although medieval pottery came from underlying silts, suggesting an earlier origin for production (NF2, appendix 3).

The saltern on the River Wyre exploited the Preesall salt works, subsequently becoming a chemical works during WW1 (NW1, 111). Presumed medieval salterns are recorded from Barrow-in-Furness and Millom, including a

Figure 4.29
Lymington, Hampshire. The 18th-century boiling house of a saltern at Creek Cottage.

possible sleeching mound at the latter (NW1, 164). Medieval salt works are common in Cumbria, in some cases being associated with monastic institutions (Wetheral Priory, Lanercost, Carlisle Cathedral, and St Bees) (NW1, 194). However, visible structures are of post-medieval date, including the site at Allonby, Crosscanonby, used from 1694–1790 (NW1, 196). A sea-water storage tank and brine pond survive (Fig 4.30). It is a rare example of a direct boiling site.

Other coastal products included kelp – used as fertiliser and in the glass industry – and wildfowl. Rutways on a shore platform at Robin Hood's Bay are though to have been associated with kelp collection for manure and potash production: the shore is still covered with *Laminaria* (YL5, 42). On Scilly, kelp burning was introduced by the Nance family of Falmouth in 1684, the soda ash produced being shipped to Bristol and Gloucester for use in glass, soap and bleach production (Fig 4.31). The industry ended in 1835. Kelp pits occur at numerous locations (IoS, 106). Duck decoys are recorded from all around the country, from Hayling Island (SE1, 20) to Hale (NW3, fig 40). Despite extensive reclamation in the Friskney area, two duck decoys survived into the 19th century, though in the post-medieval period the fowling marshes were economically significant, at least 30,000 wildfowl being sent to London annually (YL4, 28). In Norfolk, 11 decoys were recorded within the coastal zone, the most complete example being at Winterton, where a six-armed

Figure 4.30
Crosscanonby, Cumbria.
A post-medieval saltern.
The large circular feature,
c 18m across, has been
interpreted as a salt pan
or a horse gin, powering a
pump to provide sea water.

Figure 4.31
Tean, Isles of Scilly. Ruin
of a kelp-burner's cottage.

pond, 2 acres in area, was built by George Skelton in 1807, and abandoned in 1875 (N2, 152–3). In Suffolk a duck decoy is recorded at Iken, with wooden components of a decoy at Benacre (S3, 14). In Essex, Mersea was a wildfowling resort in the 17th century.

Coast specific sites: extractive industries

Mineral resources have been extracted from the source rock or alluvial sources at many coastal sites, since open exposures were convenient for quarrying, and because the products could readily be transported from them by sea.

Colliery sites of post-medieval to 20th-century date were recorded by aerial survey in the North-East, notably at the southern end of the coalfield, including the Easington and Vane Tempest collieries. Mining in Tyne and Wear, County Durham and South Northumberland extended progressively from west to east, following the dip of the Coal Measures, so that many coastal collieries are among the latest, of late 19th-century date. However, around Blyth deep coal was being mined from the 1790s onwards (NE1, 181–2). Underground workings, however, are likely to become exposed by cliff erosion (NE1, 134). There was also some smaller scale mining around Beadnell and further north, even on Holy Island, but reserves were not sufficient to supply the lime-burning industry there and imports of coal were required by the mid-19th century (NE1, 223). Besides piers, docks, quays and jetties of various types, a specialist type of structure in the North-East was the coal staithe, typically consisting of timber jetties at two or more levels. Wagons on the upper level discharged into collier vessels moored below and alongside (NE2, 9). Little remains of the industry on the ground today owing to demolition, levelling and landscaping since it ceased production (NE4, 23) (Fig 4.32).

Coal was extracted in the North-West by St Bees monastery in the medieval period (NW1, 56). Later collieries and other coal workings are known from Bees Head northwards, ranging from post-medieval to modern in date (NW3, 24–6). Denna colliery on the Wirral operated from the 1600s with larger scale extraction from the 18th century. The coal was initially shipped by underground canals to Denhall Quay. In the 19th century deeper seams were exploited, with a rail link. Post-medieval collieries focus around Whitehaven, Workington and Maryport, beginning around 1800 (NW1, 196). The Saltom Bay colliery was one where mining extended offshore. The earliest shaft was of 1729. A rock platform, 6m above high tide, housed the winding engine house and other facilities, though some buildings are now obscured by cliff collapse. Surviving structures are now a Scheduled Monument, but are threatened with erosion (NW1, 196). Collieries and waste tips were also mapped at Chislet Colliery, Kent, while Betteshanger Colliery was mapped from aerial photography as a polygon to include spoil heaps, railways and a housing estate (SE1, 49). Coal from the Forest of Dean was exported from Lydney (S1, 26).

A second extractive industry in the North-East, at Loftus, Kettleness, Sandsend and Boulby, and on the Yorkshire coast between Whitby and Reighton, was alum production, an essential fixative in the dyeing industry. It was extracted from shales, which were then calcined, extraction taking place in alum houses. The physical remains of all stages of the industry, from quarrying through processing to export all survive (NE4, 16 and 23; YL1, 99). The Boulby works originated in the 1650s, expanding in the 18th century, continuing in production until 1871 (NE2, 39; NE4, 23). Two

Figure 4.32
Cambois, Northumberland. Coastal collieries are memorialised by placing pit head wheels and trucks on their sites, but little else survives on the ground.

areas of quarrying have been distinguished, with an alum house. An expansion of 1784 led to the opening of the New Works at Loftus, where there are mounds of waste shale representing calcining clamps. Remains of steeping pits are visible in the cliff face. The alum house at Hummersea Bank, one of three serving the Boulby/Loftus quarries, partly survives as a row of four arches, but has been damaged by a landslip (NE2, 39–40, NE5, 58–63). Kettleness operated from 1727–1871. Shale processing was undertaken within the quarry, where calcining sites, steeping pits, conduits, various buildings, tracks and spoil heaps survive. The alum house has been demolished but workers' housing survives (Jecock et al 2003). Sandsend was in use between 1733 and the 1870s. Extraction of alum from shale at Saltwick Bay began in 1649 and although the alum works closed in 1791 some extraction continued into the 19th century. (YL1, 37–8). The Peak Alum Works at Stainton Dale, a scheduled site, dating from c 1650–1862, shows a well-preserved range of structures and buildings associated with the industry (YL1, 49–51). Further evidence for alum works in the area was recorded during field survey, as well as hollow ways giving access to them.

In places, the entire topography of the shoreline has been remodelled. The entire topography of Saltwick Bay was modified by the quarries, infrastructure and buildings of the alum works, with landing stages, rutways and other structures on the foreshore. Foreshore structures such as processing buildings and piers are especially vulnerable to cliff falls and erosion (YL5, 88). Foreshore rutways served both the alum and ironstone industries. They survive as parallel grooves cut into bedrock, the width of a cart axle – 1.32m or 4ft 4in. (NE1, 96) – designed to transport the product to beached vessels. The defunct railway from Whitby to Redcar, originally constructed as part of the infrastructure for the industry, survives as a tunnel from Overdale Wyke and Kettleness and as embankments and a viaduct elsewhere (NE5, 47–9). Dock facilities for the industry include those at Hole Wyke and Gallihowe Quarries, and the 'Old Gut' and 'New Gut'. These served the Boulby Alum Works. They comprise cuts into the local bedrock, some with associated postholes for timber structures and some walling. They are considered to be of 18th to early 19th century date (NE2, 40). These, and other features of the industry, are at high risk of erosion.

In Dorset, alum was extracted from pyrites occurring in Eocene surface deposits or intertidal outcrops. Extraction of alum and copperas (a related product) is recorded from Brownsea Island in 1586, and later in Kimmeridge (D2, 22). Copperas was also extracted at Whitstable (Allen *et al* 2001)

Jet extraction overlapped with the alum industries. This black fossil Mesozoic wood was used in the Roman period, but was especially valued in the Victorian period for making suitably subdued jewellery following a spouse's death. Nineteenth-century drift mines for jet are common between Hummersea Bank and Rockhole Hill, between Staithes and Port Mulgrave around Runswick Bay to Kettleness and from Kettleness to Deepgrove Wyke, where artificial caves at the cliff base extend for more than 3km of shoreline (NE2, 39). At High and Nigh Jetticks and Boggle Hole, Yorkshire, there is further evidence for jet extraction (YL5, 43). Whitby was a focus of manufacture.

From the Roman period onwards there is evidence for iron production around the Severn, for example at Chesters Roman villa around AD 250–400. Ores mined in the Forest of Dean may have been the raw material, the products being distributed via small ports around the region (S1, 24 and 32). In the north, there were probable Roman period bloomeries at Eskmeals, Drigg and Barn Scar (NW1, 160). Medieval bloomeries cluster around the upper

Kent in the North-West, representing early iron production (NW1, 136) with further examples at Sandscale Haws and Millom Castle (NW1, 164). The monks of Furness were producing iron in the 13th century (NW1, 56) and subsequently iron ore mining and processing was a major industry of the Furness peninsula (NW1, 166). Ironstone was also extracted around the Duddon Estuary (NW3, 24–6). The early post-medieval iron industry between Lancaster and Kendall was supplied with ore from the Furness peninsula by sea (NW1, 57). In the North-East the ironstone mine at Huntcliff is a Scheduled Monument but is at moderate risk of erosion (NE2, 38). The main seam of the Cleveland ironstone was discovered in 1850, leading to extensive extraction, often at the same sites as the alum shales outcropped. Production peaked in 1883, but extraction continued to the1960s. The main exporting ports were at Skinningrove and Port Mulgrave (NE1, 94–6). A fan house for the Skinningrove ironstone mine survives in a ruinous state (NE5, 62). There was also iron production in the New Forest, as at Sowley, with associated gravel and clay pits (NF1, 6).

Tin streaming on Bodmin Moor led to discharge of sediment that constricted anchorages in the River Fowey (Parkes 2000, 11). There was also tin streaming in the Helford Estuary, notably at Polpenwith (Reynolds 2000, 10). Wheal Anna Maria was mined for copper from 1833 and one shaft is visible (Reynolds 2000, 25). There was mining and quarrying for copper at Heald Brow and near Bardsea during the Industrial Revolution, with a surviving chimney at Jenny Brown's Point, Silverdale (NW1, 141; Fig 4.33).

At Filey Brigg, post-medieval and 19th-century (and probably medieval) quarrying of the local sandstones, gritstones and limestones for construction was on a massive scale and the topography of the headland was completely modified (Fig 4.34). Two mole-like structures known as The Spittalls appear to have been associated with this industry (YL1, 78–9). Surviving carved graffiti dates on quarry faces are of 19th-century date (YL5, 74). Lime production, jet working, potash mining, quarrying for building stone and clay for brickmaking were significant at several locations on this coast (YL1, 98), but field survey showed extraction of building stone to be more extensive than previously realised. Small-scale quarrying for local use was

Figure 4.33
Jenny Brown's Point,
Lancashire. The late
18th- to early 19th-century
copper-smelting site,
including this chimney,
is vulnerable to coastal
erosion.

Figure 4.34
Filey Brigg, Yorkshire.
19th-century adits into
the seam of building stone
beneath unstable coastal
deposits of till.

widespread. Larger quarries included that south of Scarborough at White Nab, where stone was quarried for building the town's quays in the 18th century. At some locations where there was easy beach access, for example at Cloughton and Burniston, stone from the foreshore was collected systematically.

The trade in Bembridge limestone originated in the Roman period, being used for most villas on the Isle of Wight and at Fishbourne Roman palace, as well as in the medieval Quarr Abbey, Carisbrooke Castle, Winchester Cathedral and the walls of Southampton, with some longer distance trade to Kent and London (Loader *et al* 1997, 23). There was some extraction from beach reefs, extending inland by 'drifts', but there were also larger inland quarries at Binstead (Tomalin *et al* 2012, 292–4). Quarries in Dorset extracted limestone at Portland, used by Inigo Jones in the construction of the Banqueting Hall in Whitehall (D2, 24). Helford was a shipping port for Carnmellis granite, providing a shipping point for quarries to the north of Constantine (Reynolds 2000, 14). Granite was also shipped from the 1830s Higher Quay at Porth Navas, where a lime kiln was also constructed. There was granite extraction at Ghyll Scaur, with other pits for sand, clay and gravel (NW1, 57). The gypsum and alabaster mine at Barrowmouth opened in the late 18th century. It expanded to comprise a range of structures with an inclined plane to transport raw material to the cliff top (NW1, 198).

Lime was significant as a component of mortar and as an agricultural fertiliser. A Roman lime kiln has been recorded at Shellness Point, Isle of Sheppey, burning seashell as the raw material (NK6, 57). Lime kilns were common on the north-east and Holderness coast from the 18th century onwards, with examples at Flamborough, Bridlington, and in rural locations (NE2, 9; YL2, 191). A late 15th/early 16th-century lime kiln was excavated at St Ebba's Nook, Beadnell, but most surviving examples are more recent (NE1, 220). Among the best preserved, albeit partly eroded, examples are of late 18th–early 19th century date at Beadnell and the Kennedy and Castle Point Limeworks on Holy Island, dating to the later 19th century (NE1, 225–6). The latter are associated with a stone pier, actively eroding. An eroding limestone kiln on the cliff edge was recorded during field survey at Scremerston, Northumberland (NE5, 194). Lime kilns and an associated ruined cottage were recorded at Hawsker-cum-Stainacre (YL5, 40) and between Gristhorpe and Filey limestone was extracted from the lower part of the cliff, beneath glacial till, for lime production (YL5, 89). Several lime kilns are known from the Helford Estuary, notably at Treath, which may have originated in the 16th century (Reynolds 2000, 12). In north Devon post-medieval lime kilns at Buck's Mills, Lynmouth and elsewhere relied on limestone and coal imported from South Wales (Fig 4.35). In the Severn, there are post-

Figure 4.35
Buck's Mills, Devon.
Lime kiln.

medieval lime kilns at Lydney and Woodend (S1, 26). Limestone quarries were recorded from aerial photography in the North-West, usually associated with kilns, as at Scales (MW1, 57). In North Kent, 19th-century and later cement works survive, notably at Elmley (NK3, 12). The Portland Cement Works and its foreshore docks, wharves, piers etc are likely to be built over for new container port facilities (NK6, 72). A cement works at West Medina Mill has been recorded (IoW, 118).

In the post-medieval period sand was extracted from the Fowey for soil conditioning on agricultural land and there were lime kilns around the estuary in the 18th and 19th centuries, using limestone from the Plymouth area and Welsh coal, as at Readmoney (Parkes 2000, 15). Sanding ways on the foreshore perpetuate routes (Parkes 2000, 26).

Clay was extracted for brick, tile and ceramic production at many localities. Most kilns recorded during the surveys were of post-medieval date, but earlier examples are known, for example in North Kent, where Iron Age to Romano-British pottery kilns occur at many places in the Medway Estuary (NK5, appendix 2). In Norfolk, brickworks were widely distributed, a good example being the West Caister brickworks on the River Bure, which reputedly supplied bricks for the late medieval Caister Castle. One long, thin pinkish-red brick of the type used at the castle was reported from the site in 1979. Aerial photographs show the site consisting of a group of large shallow pits, with platforms and banks (N2, 163–4). At Hollesley there is archaeological evidence for 13th-century pottery production, with a possible brickmaking site at Benacre (S3, 14). Associated with Quarr Abbey there was a roof-tile kiln (excavated before erosion in 1993), dated to between 1274 and 1317 by archaeomagnetism (Loader *et al* 1997, 24).

Along the Lincolnshire coast there was little industry in the 19th century, apart from fishing, although there were brickworks at Theddlethorpe and Anderby (YL7, 51). In Suffolk, brickworks are recorded on the 1881 OS map, and features representing kilns and other structures were recorded from aerial photographs at Kirton (S4, 107–10). Reede's Brickworks on the Alde was provided with a multi-phase jetty, with associated rail tracks and paving. The area is rich in brick rubble and 19th-century pottery (S1, 13). In Essex, Havengore Creek, linking the Thames and

Figure 4.36
Pitt's Deep, Hampshire. Brick wasters litter the shore adjacent to a landing designed for brick export.

Roach was used by barges taking bricks from the Wakering Brickfields to London, returning laden with street refuse and dung for use as manure (E1, 4). In North Kent 19th-century brickworks are recorded at Milton Creek and Conyer Creek (NK5, 46; NK6, 57). The brickfield at Little Cliffsend, Ramsgate, consisted of linear extraction features with banks and hollows (SE1, 47–8), and there were brickworks on Hayling Island (SE1, 19). The Newtown and Medina Estuaries on the Isle of Wight include sites of two brickworks, for example at Brickfield Farm (IoW, 99). Brickworks on the New Forest coast included those at Rushington, Exbury and Bailey's Hard (NF3, 35). Standing structures are preserved at the late 18th–early 19th century brickworks at Pitt's Deep, including the kilns. A large depression is thought to be a clay pit. On the foreshore is the 19th-century quay, comprising a hard with chains and post alignments (NF2, 48) (Fig 4.36). The brickworks at Bailey's Hard, Beaulieu River, comprised a rectangular 18th-century kiln with flues, with a later circular beehive kiln (NF2, 56). Glass, iron, copper and bricks were produced within the Liverpool dock complex (NW1, 80). Brickworks were established using river-silt rich in China Clay at Tuckingmill Creek on the Fal, 1891–1907 (Ratcliffe 1997, 59). Clay pits have been recorded around Blackpool, with brickworks at Preston, Blackpool and Fleetwood (NW1, 111).

In Poole Harbour clay extraction provided the raw material for potteries from the Iron Age to the present day with specialised production and extraction for clay pipes from the early 17th century, brickworks, and the development of Poole Pottery in the 20th (Dyer and Darville 2010, 122–35). China Clay was exported via specialised docks north of Fowey from 1869. The 19th-century jetties and mooring dolphins are still in use. They replaced shallower earlier harbours at Charlestown, Par and Pentewan (Parkes 2000, 17 and 35).

Tidal flow was first used as an energy source for milling. The earliest recorded tide mill appears to have been that at Dover, recorded in Domesday. The mill at Woodbridge, Suffolk, dates originally from 1170, though the present building is of 1790. Tide-mills were most frequent in Cornwall, Devon, Hampshire, the Isle of Wight, Sussex, Essex and Suffolk. The examples around the Solent were largely of 18th-century date, primarily for flour-milling, but there was also bone-milling to produce fertiliser or to grind the components of gunpowder. Tide mills were recorded during several of the surveys, for example at the head of the Fishbourne Channel, where breakwaters are still extant (SE2, 56).

A petroleum storage depot was established at Shellhaven in the 19th century as a response to disastrous fires on vessels at docks upstream in London. Oil refining began in the 1920s. The London and Thames Oil Company bought land around Thames Haven and Shell at Shellhaven, the two being amalgamated in 1970. The western end of Canvey was the site of the Occidental Oil Works (E1, 10). Onshore oil and gas production has so far been limited in the UK, but on the south side of Poole Harbour hydrocarbon reserves are extracted around Wytch Farm (Dyer and Darvill 2010, 118–21).

Chemical industries developed around Teeside, and in the North-West at Runcorn, Warrington and Widnes, in the late 19th century owing to the availability of rock-salt and the proximity of coal-fields. Initially alkali was the main product, needed by the glass and soap industries at Pilkington's at St Helen's, and Lever Brothers at Port Sunlight respectively. The chemical industry of the Widnes area developed in the mid-19th century, where alkali was produced using imported coal, salt and limestone. Tanning, soap and alum production were also significant (NW1, 84).

Locally, peat was a significant energy source, and in Suffolk some of the coastal Broads, including Minsmere, show evidence of peat-

cutting (S3, 14 and 50). At Alnmouth a ruinous elongated 'barn-like' structure is interpreted as a 19th-century storage building for imported guano from South America for use as a fertiliser (NE5, 175).

Coast-specific sites: military defence

Among the most conspicuous sites and monuments on the English coast are military structures. They represent repeated responses to external threat from various sources; not always nations but sometimes pirates, privateers and free-lance raiders. Later prehistoric promontory forts and cliff castles are known from many locations around England, though none seems to have been intended to counter external seaborne threats and, indeed, many may not have had a military function at all. Subsequently, external threats have come from *barbarii* and the Roman Empire itself (3rd–4th centuries AD), Scandinavian raiders and states (9th–12th centuries), the French repeatedly (in the Hundred Years War, the Napoleonic Wars and during the Victorian period), Catholic Europe more generally (in the 16th century), the Dutch (17th century), the Germans (1914–18 and 1939–5), and opponents during the Civil War (1642–1651), besides the occasional incursion of Barbary pirates (mainly 17th–18th centuries). No attempt will be made here to review the military history of coastal defence (*see* Murphy 2009, 111–43). Instead the focus will be on sites that are at risk from coastal change.

Roman sites in the north relate largely to Hadrian's Wall, campaigns extending into Scotland, and the defence of coasts nearby. The military features associated with the western end of Hadrian's Wall at Bowness-on-Solway and the string of defensive features – forts, milefortlets, towers, earthworks and roads along the Cumbrian coast southwards to Maryport are relatively well known, though in most cases not extensively investigated (NW1, 53–4). The fort at *Glannaventa* (Ravenglass) was the main centre for coastal defence between the 2nd–4th centuries. Earthworks and a well-preserved bath-house are still extant, but the western part of the site has been lost to coastal erosion (NW1, 160). Beckfoot cremation cemetery adjacent to Milefortlet 15 has been eroding for over 100 years. Erosion is at around 0.3m per year and cremations are still being recorded in the sandy cliff-face

(NW1, 191). The sites of Ravenglass and Beckfoot are considered further in NW4. Maryport, built *c* AD 122, includes a parade ground, an eroding building, possibly a temple, and other finds from the cliff-face (NW1, 188). Milefortlet 15, south of Beckfoot, is also eroding, as is 17 to the north of Allonby Bay (NW1, 189). A temporary camp at Knockcross, Bowness-on-Solway, is similarly at risk from erosion: part of the site has been lost.

Roman sites in the North-East included the signal station at Goldsborough and Huntcliff and the fort at South Shields (NE4, 22). Goldsborough signal station is 500m from the cliff-edge and at high risk of erosion. It survives as a low platform, about 40m across, originally consisting of a central stone tower with a surrounding wall and ditch. It was in use *c* 368–95 and skeletal remains have been taken to imply a violent end (NE2, 42). Huntcliff was excavated in 1911–12, when only half of it survived. It was completely destroyed by erosion in the period 1953–79 (NE1, 100). The South Shields fort, *Arbeia*, is on a low hill overlooking the Tyne. It was founded around 129 and rebuilt in stone *c* AD 160, and the site was further modified in the 3rd century as a supply base for garrison troops and operations to the north, including 18 granaries (NE2, 33). An inscription recording Tigris river boatmen at the site, presumably crewing lighters, implies that there must have been port facilities (NE1, 137). The road known as the Devil's Causeway extends to Tweedmouth; although it is principally inland, a crossing of the Tweed is implied (NE1, 211). In Yorkshire a Roman signal station at Carr Naze, Filey Brigg, was partly excavated before loss to landslip (YL1, 76). At Scarborough there is evidence for a late Bronze Age/early Iron Age settlement on the Castle headland, possibly related to a promontory fort. A 4th-century signal station was constructed there (YL1, 59).

On the east and south coasts between Brancaster, Norfolk, and Portchester, Hampshire, a string of at least ten forts of the Saxon Shore were built from the mid-3rd century onwards (Fig 4.37). Their topographic setting has changed markedly since construction. At least one, Walton Castle, Suffolk, has been completely destroyed by cliff erosion: masonry remains were visible in an aerial photograph of 1974 some 125m from the modern shoreline (S4, 43–7: *see* Fig 1.3). In Essex, coring shows that the Roman fort of Othona (now also part-

Figure 4.37
Burgh Castle, Suffolk.
A bastion of the late
Roman Saxon Shore fort.

destroyed by erosion) was originally on a promontory with creek systems extending inland; the southern one was adjacent to the fort, potentially providing a small haven (Wilkinson and Murphy 1995, 195–6). The fort at Reculver was originally sited at the mouth of the Wantsum Channel, which became infilled in the Middle Ages: it now sits on a partially eroding cliff and around half of the fort has been lost (Hunt 2011, 23–7). The shoreline has been stabilised by a ragstone apron at the foot of the cliff, which will require long-term maintenance. Excavations at Portchester Castle, Hampshire, demonstrate initial construction around AD 285–90 with ordered occupation continuing to AD 345–64 (Cunliffe 1975) (Fig 4.38).

Direct evidence for Anglo-Saxon coastal fortifications – burhs – is relatively slight, although large-scale earthworks survive at Wareham, and at Portchester Castle the Watergate Arch was rebuilt after 904. It is in a location vulnerable to marine flooding. However, later medieval castles and the defences of towns survive extensively (Murphy 2009, 121–4). Bamburgh Castle is now separated from the shore by sand dunes but originally had a coastal frontage. The 12th-century keep is the main surviving medieval feature, though an 11th-century gatehouse and three baileys are extant. There was much rebuilding in the 18th and 19th

Figure 4.38
Portchester, Hampshire.
Portchester is one of the
best-surviving late Roman
forts in the empire. It was
subsequently reused as
a medieval castle.

centuries (NE2, 18). Low-lying parts of the site are potentially vulnerable to erosion. At Scarborough there are saga references to 'Skarthi's stronghold' and there is limited archaeological evidence for 10th-century and possibly earlier activity on the headland and in the town itself. Construction of the castle was begun by William le Gros in the 12th century and it continued in use until its partial demolition after the Civil War, seeing action in that conflict and the 16th-century Pilgrimage of Grace and Wyatt's rebellion. Later 17th-century and 19th-century gun batteries were also established on the site besides a World War II RDF post (YL1, 59–67). Thus, this location retained a military significance for at least 1600 years. Currently the headland is protected by a road, though increased over-topping during storms could result in reactivation of landslips (Hunt 2011, 45–8) (Fig 4.39). The motte and bailey castle at Aldingham developed from an earlier 12th-century ringwork. Parts of the site have been lost to erosion (NW1, 135). Piel Castle, founded in the early 14th century to defend the approach to Barrow-in-Furness, comprises a keep, inner and outer bailey, moats, curtain walls and towers. Parts of the keep and south and east walls of the inner ward have

been eroded (NW1, 162). Besides castles, smaller fortifications include tower houses at Cresswell, Craster and Coquet Island (NE1, 177–8), rectangular low crenellated structures, ranging in size from around 12m to 6m. There are other medieval towers in various states of preservation at Beadnell, North Sunderland and Inner Farne – Prior Castell's Tower (NE1, 217), some of which are at risk from erosion.

The artillery forts constructed by Henry VIII, initially in 1539–43, are massive structures, generally well preserved, although parts of the outer defences of Sandgate Castle, Kent, have been eroded away (SE1, 47). Hurst Castle, Hampshire, was constructed in 1544, and subsequently modified, with major refortification in the early 19th century and again in the 1870s, when the east and west wings were built (NF2, 41–2). It is located at the end of a shingle spit which is known, from historic mapping, to have changed in form considerably over the last 150 years. The spit is at present maintained by beach recharge, which is likely to continue for the foreseeable future (Hunt 2011, 41–4), though localised erosion continues. However, there were also smaller earth and timber structures, which today survive only as earthworks. For example, at

Figure 4.39
Scarborough, Yorkshire.
The instability of the slope
up to the medieval castle
from the shore is evident.

Cudmore Grove, East Mersea, Essex, a small triangular blockhouse of 1543 is defined by low earthwork banks, currently suffering marine erosion at their seaward side (Fig 4.40). It was refurbished in 1588, as part of the Armada defences, and was again in use by the Parliamentary army during the siege of Colchester in 1648. Excavations on the eroding foreshore have revealed wooden structures, including a timber quay, and probable beacon (E Heppell, pers comm). On the North Norfolk coast at Cley-next-the-Sea, a raised area on the salt-marsh surrounded by a bank and ditch has been provisionally interpreted as the remains

Figure 4.40
East Mersea, Essex. The
low earthworks of the
Henrician fortification
are being eroded.

of Black Joy Forte, an Armada-period fort recorded on a map of 1588. Despite later changes at the site, the NMP defined an angular embankment, some 265 by at least 65m, which could represent this structure (N2, 155–6; N1, 151).

The strategic situation of the Isles of Scilly, well placed to control contact between Ireland or Scotland and France or Spain, led to the development of a complex of military defences from the mid-16th century onwards. The islands were also the scene of combat during the Civil War; while new military facilities related to air forces were established in World Wars I and II. The surviving structures exemplify changes in military defence over some four centuries (Bowden and Brodie 2011; IoS 79–85). The earliest defences, constructed in the reigns of Edward VI and Mary I (1547–58), comprised blockhouses and forts (King Charles' Castle, Harry's Walls) on Tresco and St Mary's, though Harry's Walls was left incomplete. Later structures include Star Castle (1594) and the earliest phase of the Garrison Walls. At the beginning of the Civil War the islands were held by the Royalists, who were dependent on privateering for their support; the Parliamentarian assault came in 1651. A Dutch engagement on the islands, in response to a capture of their shipping, was repulsed by Blake in the same year. Construction in this

period was mainly of earthwork batteries (typically V-shaped and sited on cliffs, though at some sites more elaborate, including Oliver's Battery on Tresco) and breastworks. The circular blockhouse known as Cromwell's Castle was a stone structure, built in the 1650s. There was some new construction following Lilly's survey of the islands' defences in 1715 and major extension of the Garrison Walls during conflicts between Britain and Spain from 1739 onwards, including the War of Austrian succession (1740–8). Construction of the walls, incorporating redans and bastions, ceased in 1747, though they were reactivated during the Napoleonic Wars, with some modification of bastions to accommodate more modern ordnance and construction of gun towers and a telegraph tower. In the 19th century, construction involved new batteries – the Woolpack and Steval Batteries – to house breech-loading guns, housed in concrete emplacements masked with earthworks, besides new barracks and searchlight batteries.

Management of the outer Garrison Walls is complicated by the geology of St Mary's, which consists of a layer of stone and clay 1–3m thick, known as 'ram', over granite bedrock. The ram is susceptible to undercutting by wave action, especially during storms, leading to shoreline retreat (Fig 4.41). Parts of the walls have been reconstructed following collapse and the

Figure 4.41
Garrison Walls, St Mary's, Isles of Scilly. This structure has been repaired but is in a precarious location.

shoreline at other locations will require armouring of the shore locally (Hunt 2011, 32–6).

In the 18th century threats to the north-east coast led to the development of 'defended ports' with new batteries at existing sites, including the North Battery on Hartlepool Headland (NE2, 10, 36) and at Tynemouth (NE1, 152). During the 19th century additional gun batteries were developed at Fairy Cove, The Heugh and Lighthouse Battery. There were other 19th-century batteries at Redcar NE2, 36–7). A late 18th-century gun battery at Whitby was sited on the West Pier, where embrasures for cannon still survive; there were three other batteries (including the partly extant site of the Half Moon Battery), which would have provided overlapping fields of fire to defend the harbour and prevent landings (YL5, 33, 87). Structures of this type are especially vulnerable to redevelopment.

On the East Anglian, Essex, Kent and Sussex coastlines Martello Towers were constructed in the early 19th century (Fig 4.42). Some are associated with batteries (at Winchelsea) and other forts, such as the triangular Fort Sutherland at Hythe, built 1798 but now demolished (SE2, 94–5). The northern end of the system, at Slaughden near Aldeburgh in Suffolk, is marked by a quatrefoil tower with four guns, basically four normal towers conjoined (S3, 15). Parts of the outer defences have been lost to erosion.

Coastal defences of the Victorian period include the Royal Commission forts of the 1860s, and subsequently low-profile earth and concrete forts for breech-loading guns, and offshore sea forts, for example at Horse Sand off Portsmouth (*see also* Fig 4.43). Portsmouth dockyard and the associated Fort Blockhouse, Gosport and the Round and Square Towers were mapped and their positions corrected during survey (SE1, 17). Other military sites mapped included Fort Monckton, Fort Gilkicker, Gosport Lines and Bastion No. 1, Gosport town moat, Batteries 1–5 and the

Figure 4.42
Folkestone, Kent. Martello Tower during conversion for residential use.

Figure 4.43
Fort Brockhurst, Gosport, Hampshire. One of many defensive features around Portsmouth Harbour.

Stokes Bay Lines, Haslar Gunboat Yard, Browndown Battery, Priddy's Hard, Royal Clarence Yard and the Haslar Royal Naval Hospital (SE1, 17–18). The Old Needles Battery was associated with searchlight experiments in the 1890s (IoW, 77). Warden Point battery is associated with military road and marker stones and on the sea wall below Warden Point battery is a late 19th-century searchlight base (IoW, 81). Further north-east is the Cliff End battery and Grade II* Fort Albert, with further searchlight emplacements (IoW, 83). Further on was Fort Victoria, on the site of previous defences. Again, emplacements for searchlights focused on the Needles Passage (IoW, 87).

Since the conflict of World War I took place principally overseas, there are relatively few surviving sites of the period in this country; indeed reuse of the same sites in World War II often resulted in destruction of earlier facilities. Batteries were developed around strategic locations, for example at St Anthony Head and Half Moon Battery covering the Carrick Roads to Plymouth (Ratcliffe 1997, 29–31). Attacks by the German High Seas Fleet on Whitby and Hartlepool in 1914 provided the impetus for construction of new batteries, for example the Coulson Battery, South Beach, Blyth, which provided additional protection both for the coal port of Blyth itself, and for the Tyne (NE1, 61), and similar batteries were subsequently established near Great Yarmouth. At

Tynemouth gun turrets from HMS Illustrious were mounted on land towards the end of the war at the Kitchener and Roberts Batteries, removed by the 1920s. Nothing survives of the former, but the command posts, officers quarters, water tower and other features at the Roberts Battery survive and are listed Grade II* (NE1, 154). The main focus of military activity near the Humber comprised the Godwin Battery, Kilnsea and Spurn Fort, comprising Green Battery and Light Permanent and Light Temporary Batteries, built 1915–16. Infrastructure included barracks, searchlight emplacements, observation posts, blockhouses, Port War Signal Stations (to identify friendly shipping) and railway (YL6, 103). The Godwin Battery has now been almost totally destroyed by erosion, with most significant structures now lying collapsed on the beach (YL6, 103). Offshore, Bull Sands Fort and Haile Sands fort on the Lincolnshire side were built of concrete on steel foundations in 1915 (YL2, 192–3). There are relatively few surviving anti-invasion defences, although in Lincolnshire, rectangular blockhouses survive at Skidbrooke and possibly North Somercotes (YL7, 51), with some circular examples with an overlapping roof in eastern England and Kent, whose appearance coined the term 'pillbox' (Foot 2006, 2). In Yorkshire and Lincolnshire small square pillboxes and a blockhouse dating from World War I have been recorded (Brigham et al 2012) (Fig 4.44).

Figure 4.44
Skipsea, Yorkshire. Possible forward command/ communications post SK49, dating from World War I, displaced and deteriorating.

Figure 4.45
World War I practice
trenches, Gosport
(Hants MWX60383).
(RAF/540/453 RS fr.4185
© English Heritage,
RAF Photography)

Training facilities for the Western Front included firing ranges and practice trenches, with their characteristically crenellated ground plan, plainly visible in aerial photographs, though now infilled. Ranges were recorded at Barrow-in-Furness and Fleetwood, with practice trenches at Blackpool (NW3, 27–35). In Norfolk there were the crenellated practice trench systems on Kelling Heath, and firing ranges at West Runton (N2, 165–70). Surviving World War I military features are not common in the North-East, but included a practice trench system comprising a front-line trench, reserve trenches and communication trenches with saps at Saltburn Marske and New Marske (NE4, 24). On Levington Heath, Suffolk, crenellated trench systems are also thought to be of this period (S4, 122–5), as are other sites in the South-East (Fig 4.45).

This was, of course, the first war involving aerial combat. There was a coastal Anti-Aircraft battery at Barrow (NW1, 167). At North Gare, Teeside, the slipway, pier and associated buildings of a seaplane base were recorded during field survey (NE5, 87–8). The jetty and slipway for the Seaton Carew seaplane station survive (NE1, 110). At North Cotes on the south Humber bank there was an airfield in 1916–19, subsequently reused in World War II, and from 1957 a bloodhound AA Missile base, up until 1990 (YL2, 166). On the Holderness coast, sites include the airfield at Atwick and Royal Naval Air Service Seaplane base at Hornsea Mere. In Norfolk the South Denes seaplane station (with prominent seaplane sheds) was recorded from aerial photographs (N2, 165–70). In Kent sites include the Capel-le-Ferne/Folkestone Airship Station, used for anti-submarine patrols. Remains of the hangars are partly visible in arable fields (SE2, 97). The flying-boat station at Hythe and the military camp at Eaglehurst, Calshot, were recorded in detail from aerial photographs. It accommodated servicemen serving at the Calshot flying-boat station (NF3, 38). The Isles of Scilly were of significance principally as a flying-boat and seaplane base at Porthmellon and, later, New Grimsby, Tresco, to patrol the Western Approaches in search of U-boats and surface vessels (Bowden and Brodie 2011).

At Portland a breakwater was constructed in 1872, followed by others in 1894, which led to the site becoming a habitual anchorage for the Channel Fleet, where the Reserve Fleet and Grand Fleet assembled in 1914, before dispersing to war stations (D2, 12–13). Richborough Port, constructed in 1916 to supply the military in France, covered some 4.8 × 2km, including a wharf, shipyards, railways, barracks and storage sheds. It became disused in the 1920s, but was partly reused in World War II (SE1, 49; SE3, 81). In fact, World War I sites frequently continued as defensive sites later, as at Orfordness, Landguard Fort, and RAF Bawdsey. Orfordness was the research base for a Royal Flying Corps base, 1915–1921, associated with bomb stores, a World War 1 prisoner of war camp and a motor transport shed. It later became an experimental firing and bombing range, 1921–1939, and was the first base for Watson-Watt's experimental radar base, from 1936, expanding to RAF Bawdsey. Concrete bases of receiver masts are visible. The site is being eroded.

An inter-war innovation was the use of concrete sound mirrors to detect incoming aircraft. In Kent sound mirrors were emplaced, dating from the 1920s–1930s at Abbot's Cliff, Hythe and Denge. They were abandoned in the mid-1930s, being replaced by radar (SE2, 98–9). In the North-East the sound mirror at Boulby Barns is one of the few surviving (NE1, 110) and there is also a sound mirror at Kilnsea (YL6, 103).

The anti-invasion defences of England, built mainly in 1940 during World War II, have been reviewed by Foot (2006). Despite this authoritative study, the RCZAS have produced still more information, partly from aerial photography and ground survey – and not just about defence, but also training and offensive facilities. In the Portsmouth and Hampshire areas aerial photographic mapping added 638 records and enhanced 482, resulting in more than a doubling of records for Hampshire and an increase of 150% for Portsmouth. Of these, the vast majority dated to World War II: 186 new records and 67 enhanced records for Hampshire; 181 new records and 1 enhanced record for Portsmouth (SE1, 14–15). In East Kent – then known as 'Hellfire corner' – over 60 per cent of the HBSMR records created or enhanced for the National Mapping Programme in Blocks M and L were of this period (SE1, 50). Given both studies, we may be close to having a full record of sites, which were not always documented at the time, but it is not possible to discuss it at all fully here. Instead a number of key sites, exemplifying World War II coastal military structures will be considered. Being directly on the coast many of these structures have been, or are being, destroyed by coastal change (Fig 4.46).

The 'coastal crust' of defence comprised coastal batteries and 'hardened field defences' consisting of weapon emplacements and concrete anti-tank obstacles (Fig 4.47), as well as minefields, anti-tank scaffolding, anti-tank

Figure 4.46
Kilnsea, East Yorkshire.
A collapsing World War II defence opposite the Crown and Anchor.

Figure 4.47
Walberswick, Suffolk. Anti-
tank defences relocated to
form parts of sea defences.

ditches, fire trenches, weapon pits, and other earthworks, and barbed wire entanglements' (Foot 2006, 8). Other features included searchlight emplacements, radar stations, bombing range markers, bombing decoys, air raid shelters, practice trenches, and barrage balloon sites. Anti-glider ditches and poles were recorded on sand flats in Morecambe Bay and adjacent to the Anglo-Saxon barrows around Sutton Hoo. Many of the Heavy AA batteries were related to the DIVER defence against early guided missiles from 1944. Existing defensive sites were also remodelled, as at Landguard Fort, where the 1540s defences and their modifications up to 1875 were refortified in World War II, with new guns and searchlights (S4, 125–38). Along most of the coasts of southern and eastern England, more or less continuous defences were constructed from 1940–42, with gaps at cliff locations. Pett Level was flooded by breaching the sea defences, while Dymchurch was provided with a concrete sea wall and anti-glider defences. The 19th-century Royal Military Canal was incorporated into 20th-century defence (SE2, 101–17). This coastal defence was linked to defence in depth, initially by a stop-line, a prepared battle-field.

A comprehensive survey of World War II military remains within the Northumberland AONB (Area of Outstanding Natural Beauty) was undertaken, partly for record, but also to inform future management (NE5, 237–69). The entire area of the AONB, 138km², was covered. As elsewhere, pillboxes of various forms were emplaced at weak points in the coastal defence, and were usually associated with other defensive components, such as trenches and weapon pits, which survive less well. Well-preserved examples survive in dune systems between Dunstanburgh and Embleton. Anti-tank blocks survive in several areas, though inspection on the ground showed that some had been removed, often serving as sea defences, while others were partly buried by modern dune systems. Gun emplacements consisted of coastal batteries and machine-gun emplacements. Coastal batteries at Goswick and Budle housed 6-inch guns. Machine-gun emplacements, distinguished from pillboxes by their wider openings, covered potential landing beaches, as at Beadnell Bay and Bamburgh. The only airfield, at RAF Boulmer, had been used as a bombing decoy, before 1943, for RAF Acklington, and so air-raid shelters were supplied. Other features recorded comprised earthwork remains of minefields, anti-glider ditches and poles.

Druridge Bay was potentially an ideal invasion beach and hence was developed as a Defence Area. Pillboxes and anti-tank blocks were built in the dunes, with scaffolding, ditches and minefields on the beach, and with observation posts and anti-glider ditches in fields behind. A pillbox was built, for camouflage, within the 14th-century Chibburn Preceptory and another is disguised as a ruined cottage (NE1, 195 and fig 8.23). Other sand beaches, as at Sandsend, North Yorkshire, were likewise defended: anti-tank blocks and a collapsing pillbox were recorded during field survey (NE5, 50–2). At Crimdon Dene, pillboxes, anti-tank blocks and an anti-tank battery were recorded (NE5, 99–103) and at Trow Point, Tyne and Wear, a complex of partly collapsed and eroding World War II features was recorded (NE5, 111–16). On the cliffs further north the site of a World War II camp was examined, producing the well-preserved remains of a Vickers anti-aircraft machine gun post (NE5, 53).

The World War II defences of Teesmouth, Wearmouth and Tynemouth Defended Port were mapped in detail from aerial photography during the RCZAS. They comprised a complex of anti-invasion defences, air defences, obstruction sites and military training/accommodation sites (NE1, figs 6.13, 7.24–5). Further south, at Greatham Creek and North Gare, Teeside field survey recorded surviving features ranging from earthworks such as anti-glider ditches to standing buildings and structures (pillboxes, section posts, spigot mortar bases, anti-tank blocks (NE5, 80–5). There is also a concentration of surviving World War II defensive structures around Freiston, Lincolnshire, at the mouth of the River Witham, including the Freiston Shore battery, comprising 6-inch gun emplacements, pillboxes and searchlight installations (YL4, 41). Another heavily militarised area was around Portsmouth, where anti-tank obstructions included lines of defences adjacent to Fort Cumberland (SE1, 29). Other establishments in this area included Heavy Anti-Aircraft Batteries on Southsea Common, Hayling Island, and northwest of Fort Brockhurst (SE1, 30), barrage balloon sites, searchlights, and anti-landing obstructions.

Decoys were developed at many locations to divert bombers from industrial targets and ports. They included QL decoys, with lighting to mimic ineffective black-out and SF (Starfish) batteries to mimic fires started by bombs. There was an extensive bomb decoy site on Farlington Marshes, next to Portsmouth. Brownsea Island was a Starfish Decoy, which still survives remarkably intact, with a control bunker, store, bomb shelter and, tellingly, numerous bomb craters (Dyer and Darvill 2010, 100–1), while there were others on Beaulieu Heath in the New Forest. Numerous bomb craters were recorded, along with shelters and water storage tanks (NF3, 38–48). In the North-West there was a decoy site at Ravenglass to help protect Vickers-Hycemore (NW1, 167) and another between the Maryport and Whitehaven, to protect the Workington Iron and Steelworks (NW1, 199).

A major defensive system was established in the North-West to protect the shipyards, including RAF Millom and Walney Island, with pillboxes, coastal batteries, gun emplacements, trenches and weapons pits, AA (Anti Aircraft) Searchlight batteries, decoys, barrage balloon sites, obstructions, road blocks, camps, firing ranges, bombing range markers, shelters and emergency water supplies (NW1, 167). There were two coastal AA batteries on Walney – a site rapidly eroding. The Royal Ordnance Factory at Sellafield has been replaced by the nuclear facility (NW1, 168). In general the North-West appears to have been used extensively for training, construction and defensive purposes, being further from enemy airfields. Liverpool, however, was heavily bombed in World War II: over 90 per cent of imports went through the Liverpool docks (NW1, 84).

Structures associated with the invasion of Europe (Operation Overlord) in 1944 include the production and embarkation sites for Mulberry Harbours (Figs 4.48–4.49), for example at Marchwood military port, which developed from a small village (NF3, 38–48), at Southampton and Langstone Harbour (SE2, 77–8), and Stokes Bay, and Sinah Warren, Hayling Island, where an imperfect caisson still survives (SE1, 30). At Lepe the foreshore shows slipways, anchoring points, bollards, winching gear bases, trigger release gear sites, hardening mats and platforms related to the use of the site for Mulberry Harbour construction (NF2, 57). Thorness Bay on the Isle of Wight includes a landfall for Operation PLUTO (Pipe Line Under The Ocean), where a manifold survives on the shore (IoW, 104), and there were pumping

Figure 4.48
Hardway, Hampshire.
Embarkation slip, used
by Canadian forces on
10 June 1944, for invasion
of Europe.

stations (disguised as houses and a chapel) for PLUTO, north of Dungeness (SE2, 101–17). Portland also shows embarkation slips built for the American forces that attacked Omaha beach in 1944 (D2, 12–13). The D-day embarkation point at Trebah survives as a hard and pillbox (Reynolds 2000, 17–18, 22).

Civil defence air-raid shelters were mapped at most towns, for example at Great Yarmouth, some being communal, some private, largely consisting of Anderson Shelters (N2, 170–212). Civilian facilities, including air raid shelters, emergency water supplies, and allotment plots were mapped in the Portsmouth area (SE1, 31–4). After World War II, pre-fabricated homes, meant to be temporary, were constructed: in fact 'pre-fabs' remain in use up until today in some places (SE1, 31–4). Industrial sites comprised factories, such as the Spitfire factory (Supermarine Aviation Works)

Figure 4.49
Hayling Ferry, Langstone
Harbour, Hampshire.
Abandoned section of
Mulberry Harbour.

at Woolston and Fairey Aviation's seaplane factory on Hamble Point (SE2, 80–81).

In the longer term, however, the most lasting impact on the English urban landscape, however, was the terrible destruction by bombing of ports and coastal cities, mainly in 1940–1. In Southampton 30 per cent of the housing stock was destroyed. The historic cores of many major ports were directly adjacent to modern port and dockyard facilities, as at Great Yarmouth, Ipswich, Portsmouth and Southampton, and suffered accordingly. Direct evidence for bomb damage is now sparse, for extensive reconstruction has taken place, but at Long Reach, Darenth, architectural stonework of classical style derived from bomb-damaged buildings in London is strewn on the shore, representing the cargo of a sunken barge (NK6, 11). Some surviving structures, such as Fort Cumberland, Portsmouth, show masonry replaced by concrete repairs mimicking the original stonework: here the line of a stick of bombs can be seen, which killed a party of Royal Marines sheltering in a casemate in 1941. The new entrance to the fort is at their scene of death, where the rampart was destroyed.

Aircraft crash sites, mainly from World War II, have been recorded at coastal locations, including the Medmerry, Sussex Managed Realignment Scheme. Aircraft losses in the area are recorded historically, and one site thought to be of a German Ju88 has been detected

(Jonathan Sygrave, pers comm). Prisoner-of-war camps were developed at several locations, including North Lynn Farm, Norfolk, where groups of huts and an ornamental garden are visible in a 1946 aerial photograph (N2, 170–212).

Royal Observer Corps (ROC) monitoring posts, which persisted into the post-war period, survive at various locations, for example at North Gare, Teeside, where a hatch and ventilation shaft had been inserted into a medieval saltern mound (NE5, 87–8). Cold War ROC posts, for example at Meon and Dell Quay (SE2, 80–1). On the Isle of Wight the Old and New Needles Batteries survive, the latter being later used as a site for testing Black Knight and Black Arrow rockets. World War II features include an AA emplacement, site of a spigot mortar base and trenches (IoW, 75). After the war, Orfordness, Suffolk, continued in use as a research establishment (Fig 4.50) and a Bloodhound battery was established at Bawdsey (S4, 125–38).

Coast-specific sites: control, navigation, rescue

Customs, in roughly their present form, date from the reign of Henry VII. Elizabeth I tightened the system in 1558, by restricting by statute the numbers of landing places to legal quays, where goods of certain types had to be

Figure 4.50
Orfordness, Suffolk.
The 'Pagodas', used to test
conventional charges that
would have triggered
nuclear bombs.

loaded and landed and where customs could more readily be levied. At London, for example, the first legal quay was just below London Bridge. To house the administrative staff, ports of any serious standing built Customs Houses, or adapted existing buildings. They were often of considerable architectural distinction: the Customs Houses at King's Lynn (1683) and Poole (1813) are fine surviving examples (Fig 4.51).

Housing the Coastguard to monitor evasion posed problems in terms of susceptibility to corruption, bribery and social exclusion. Given that many of the locals would have been involved in smuggling, they would not have welcomed the representatives of official authority into their villages. The solution was the development of Preventative Stations, providing separate accommodation. The Coast-guard cottages at the edge of modern coastal villages, frequently simple but elegant late Regency or early Victorian buildings, still survive extensively and are noted in several of the RCZAS reports (Fig 4.52). The Coastguard Act 1856 placed control under the Admiralty.

There were medieval organisations concerned with navigation and pilotage, but Trinity House itself was established originally at Deptford, in 1514, with responsibilities for lighting, marking and buoying. Trinity Houses were also developed in Newcastle (originating from the medieval Guild of Masters and Mariners), and Kingston-upon-Hull, with comparable societies and companies at Bristol and Dover. As was the case with Customs and Excise, the dignity of office was marked by buildings of some architectural distinction, notably the imposing and pleasing structure at Kingston-upon-Hull.

Figure 4.51
Poole, Dorset. The
Customs House of 1813.

Figure 4.52
The Garrison,
St Mary's, Isles of Scilly.
Coastguard cottages.

Figure 4.53
Dover, Kent. The surviving
Roman lighthouse.

Figure 4.54
South Shields,
Northumberland.
Tynemouth Volunteer
Life Brigade look-out.

The earliest surviving lighthouse is the Roman structure at Dover (*Dubris*), originally one of two on either side of the harbour (Fig 4.53). To list and discuss all the surviving examples noted in the RCZAS is impossible here, but they often had long and exciting histories. The approach to the Tyne was especially hazardous owing to mobile

sandbanks and the reef of The Black Middens, and in the North-East one of the earliest lights was provided by the monks of Tynemouth from the monastic towers. These finally collapsed in 1659, being replaced by a new structure with a coal-fired brazier. This brazier was rebuilt in 1775 and an oil-fired light installed in 1802, finally being demolished in 1898 (NE1, 144), being replaced by lights on the North Pier and St Mary's Island. The light at Souter of 1871 was among the first to use electric light (NE1, 142). On Coquet Island the light, built 1839–41, incorporates parts of a 15th-century tower house (NE1, 189). The ruinous medieval St Helen's church tower at The Duver on the Isle of Wight now acts as a sea mark, painted white on its seaward face (IoW, 29). As a result of coastal change and reclamation, some lighthouses have ended up well inland, such as the two lighthouses at the former mouth of the Nene at Sutton Bridge, Lincolnshire (YL4, 66).

The earliest dedicated craft for life-saving is thought to have been a converted coble at Bamburgh, in the late 18th century, but development of the modern lifeboat service has its origins in 1864, when the *Stanley* was wrecked on the Black Middens, off the River Tyne. Rescue attempts by the Coast Guard were hampered by the failure of the shore-to-ship 'Life Saving Apparatus', developed by John Bell in 1791. This was basically a mortar, firing a projectile, with a line attached. The severe loss of life led to development of the Tynemouth Volunteer Life Brigade, and the idea was widely adopted (Fig 4.54). Regional provisions eventually coalesced into the formation of the entirely volunteer Royal National Lifeboat Association after 1854. Numerous splendid Victorian lifeboat houses survive, and are noted in RCZAS reports. However, some are now defunct and eroding, including the example at Hilbre Island, Merseyside.

Coast-specific sites: religious foundations.

Prehistoric and Roman religious and burial sites lie on the modern coast, but attribution of any of them to worship of a marine deity is uncertain, except perhaps in the case of the Roman-period site excavated at Nornour, Isles of Scilly. The rich artefact assemblage – coins and brooches – has been interpreted as possibly votive (Fulford *et al* 1997, 173). Bronze Age deposition of human remains and artefacts is generally interpreted as ritual or symbolic in character; the Cliff Castles of western England have been interpreted as possible Iron Age ritual sites, and the Anglo-Saxon burial site at Sutton Hoo, Suffolk, has been seen as a late assertion of paganism in the late 6th to early 7th centuries AD (Murphy 2009, 280–9). But in most cases a coastal location need not necessarily have religious meaning.

Several of the most important Anglo-Saxon monastic foundations in the North-East, from the early 7th century onwards, lie within the coastal zone: Lindisfarne, Tynemouth, Jarrow, Monkwearmouth, Hartlepool and Whitby. These Anglo-Saxon and Celtic religious establishments were founded at apparently remote coastal locations but they were then frequently next to busy sea-ways. They combined the qualities of isolation for those who did not own a boat with communication for those who did. Frequently these early foundations were replaced by later medieval monastic sites that became linked to the wider economic life of the country (NE1, 56). The monastery at Lindisfarne was a gift of Oswald of Northumbria to St Aidan in 635, and it developed until the first Viking raid of 793, being abandoned in 875. It was subsequently reconstructed in the 11th century. Excavations have revealed elements of the early monastic site, including its boundary ditch, but nothing now survives above ground (NE2, 15). At the adjacent St Cuthbert's Island, structures mainly of 13th-century date are actively eroding, perhaps along with buried archaeology related to the saint himself (NE2, 15; NE5, 216–18). The chapel is a two-roomed structure, with a related building possibly for accommodation or storage, and earthwork features include a stone-lined drain and sub-circular mound, besides a slipway. The 13th-century St Ebba's Chapel, Beadnell, is also at high risk of erosion (NE2, 18). Whitby was founded by Hilda in 675, but the present abbey was re-founded in 1078, following destruction of the earlier foundation in 9th-century Viking raids. At East Cliff, the site of the early monastery (WH25) and its associated cemetery and settlement, there has been extensive excavation in advance of cliff erosion (YL1, 29–30).

Monastic sites would have had harbours, but of these little is known. Blythburgh, Suffolk, was the burial place of King Anna in 654 and may have been a royal residence. The church itself may have been a Minster: high-status

finds include styli and a whalebone writing plaque and Middle Saxon Ipswich Ware has been found around the Blyth, potentially marking a port (S3, 43). A later example comes from the Cistercian foundation of St Leonard's Grange in the New Forest, which is represented today by a 14th-century chapel and barn and a probable harbour at Gin's Farm, Hampshire (NF1, appendix c).

Middle to late Saxon monastic sites in Suffolk include the first bishopric at *Dumnoc*, which may have been either at Dunwich or Walton Castle, both sites having been lost by erosion. The Premonstratensian Abbey, Leiston, originally of 1182, was relocated inland from its original site on a low rise in the marshes owing to later threats from sea flooding: an early example of Managed Realignment. Some architectural components from the first abbey were built into the new structure. Some ruins of the original structure survive, with clear cropmarks defining its lay-out. The site will be vulnerable in terms of proposed modern Managed Realignment (S3, 48). In Essex the early chapel of St Cedd at Bradwell is a rare surviving structure from early Anglo-Saxon Christianity, having later been converted into a barn before its modern reinstatement for worship (Fig 4.55).

Early Christianity in the North-West is relatively well understood, with excavations at St Patrick's Chapel, Heysham, Lancashire, revealing an 8th-century precursor with a

Figure 4.55
Bradwell-on-Sea, Essex. St Cedd's Chapel of 654. The early monastery was situated inside a former Roman Saxon Shore fort, perhaps to exploit recyclable Roman stone but could also reflect a reference to the glories of the Imperial past.

Figure 4.56
Heysham, Lancashire.
The 8th-century St Patrick's
Chapel, which is associated
with rock-cut graves
of indeterminate, but
perhaps earlier, date.

cemetery containing rock-cut graves (Fig 4.56). Other headland sites include the Chapel of St Hildeburgh on Hillbre Island, St Michael, Workington, and many cross fragments (NW1, 54–5). Later ecclesiastical sites in the North-West included Cockersand Premonstratensian Abbey, which survives above ground as the 13th-century Chapter House. The site sits on an eroding cliff, which has been stabilised, but the long-term future of the site is uncertain.

More isolated foundations have also been recorded. A crudely constructed cell for a solitary has been recorded at St Ebba's, Beadnell (NE1, 68) and the monastic cell on Coquet Island is dated to the 12th century (NE1, 178). Finds from Coquet Island also include a cross-incised stone slab and a 9th-century ring: there is a documentary record of a meeting between St Cuthbert and Abbess Elfleda of Whitby there in 684 (NE 1, 176). The flint-walled structure known as Blakeney Chapel, on Blakeney Freshes, proved on excavation in advance of coastal realignment to be of 15th–16th century date. The finds assemblage comprises mainly items of domestic or personal character, and it included a cobbled floor with internal partitions and a fireplace and so it is more likely to be a domestic structure than a chapel, conceivably a Warrener's Lodge (Birks 2003).

Parochial medieval churches are at risk of erosion in many places, especially in East Anglia. Almost all the churches of medieval Dunwich, Suffolk, have been lost. Covehithe is a large, partly ruinous church, reflecting both high medieval affluence and a later decline. Archaeological material from the vicinity indicates extensive occupation in the 14th–16th centuries, probably associated with a now-lost sheltered harbour within Benacre Broad (S3, 13). The coast here is among the fastest-eroding in the country and the site will be at risk later this century. In Norfolk, the village of Eccles-next-the-Sea was gradually destroyed by coastal erosion from the 17th century onwards, owing to dune recession. The church tower survived until 1895, when it collapsed. Following storms, the rubble from the church is occasionally exposed (N2, 11).

Coast-specific sites: tourism and resorts

A history of English seaside resorts is given by Brodie and Winter (2007) but the RCZAS have produced additional data. Spa water was noted at Scarborough as early as 1620 by Elizabeth Farrow (YL1, 66). Filey was a resort from 1670, also with a spa, expanding in the 19th century (YL1, 81). Whitby and other minor coastal resorts, for example Hayburn Wyke, were also developed as resorts; at the latter location modifying the landscape to provide easy access for visitors. The place-name *Tea Grounds* at Saltwick and Scalby, at former industrial sites, is perhaps evidence for diversification as extractive industries declined (YL5, 89). Railway expansion, as at Bridlington in 1846, permitted easy access from the 19th century onwards, alongside construction of the North and South Promenades, hotels and guesthouses, and a spa in 1896 (YL2, 190–1). Recreational development at Hornsea and Withernsea was also of 19th-century date, again after railway construction, in 1864 and 1854 respectively (YL2, 79, 111).

There was a hotel at Skegness by the end of the 18th century, and bathing machines were on the beach in 1784 (YL3, 55), but the extension of the railway network in the 19th century resulted, as elsewhere, in large-scale tourism at resorts such as Mablethorpe, Chapel St Leonards, Ingoldmells and Skegness, followed by the expansion of holiday camps and caravan sites in the 20th century. Skegness was developed by the 9th Earl of Scarborough in the late 1870s after the Great Northern Railway reached the town in 1873. Butlin's camp at Ingoldmells was built in 1935–6 and is still operating, but had a hiatus during World War II when it was in military use (YL3, 69–70). At Mablethorpe, railways linked Louth and Willoughby with the developing resort from 1872 and 1878 respectively, and a narrow gauge tramway connected Alford to Sutton (1884–9). Freiston Shore was developed as a resort in the 18th century, one of several around the Wash (YL4, 40). This was a failed resort. Of its buildings the Marine Hotel is ruinous, Plummer's Hotel is still in use, and the bathing house demolished (YL8, 42 and plates 54–5).

In Norfolk, Mundesley, Cromer and Great Yarmouth all became resorts during the 18th century, followed in the 19th century by New Hunstanton, Wells-next-the-Sea and Sheringham (N1, 11–12). Canvey Island in Essex was another failed resort. In 1901 a mono-rail tramway and Winter Gardens were built, then dismantled on the contractors' bankruptcy. However, residential development continued (E1, 9). The seaside resorts in Dorset often arose from existing coastal settlements and ports, though Bournemouth was a deliberate development by local landowners and the local Improvement Commissioners.

The town was planned to confine business development well to the north of the coastal zone of villas, and working-class accommodation was excluded (D2, 9). Spas are recorded at Radipole and Nottington in the 1720s, with foreshore licences for bathing machines at Weymouth in 1748 (D2, 18).

Coastal resorts expanded during the 19th century in the North-West, including Blackpool, Morecambe and Southport, to provide holidays principally for the Lancashire working class. Some developed in an unplanned fashion, such as Blackpool, but others including Fleetwood and Morecambe are among the best examples of coastal town planning in England (NW1, 58). Blackpool, Lytham and Fleetwood originated from a group of small fishing villages after the railway was constructed in 1846. Before the mid-19th century visitors came principally for bathing and drinking the water, but pier construction followed in the 1860s and 1890s. Mass entertainment was established by the later 19th century, including hotels, which could have accommodated over a quarter of a million (NW1, 111). The Blackpool Tower was constructed in 1894, with a Captive Flying Machine in 1905. Morecambe developed in the later 19th century, with piers built in 1870 and 1893, now both destroyed (NW1, 142).

Higher status coastal establishments are marked on the Isle of Wight by the estates of Norris Castle and Osborne House. At Osborne two listed (Grade II) structures – the Queen's Alcove (Fig 4.57) and an Italianate style boat-

Figure 4.57
Osborne House, Isle of Wight. The Queen's Alcove before restoration. It offers a fine view across the Solent, where Her Majesty's fleet could often be seen.

house – have recently been restored and opened for public access (IoW, 11). A large, ruinous and partly eroded bathing house marks the western boundary of the Norris Estate (IoW, 114). There are ornamental designed landscapes of 18th century and later date in the Fowey Estuary, including salt-water baths and boat-houses. A few are focused around former gentry houses, but others are later: the Rashleigh house at Menabilly from the 16th century is an example (Parkes 2000, 22).

However, not all resorts were of such high status. The philanthropist Joseph Cunningham is generally credited with establishing the first Holiday Camps as early as 1894, when he leased land on the Isle of Man for the use of workers from Liverpool. Commercial holiday camps began in the 1920s, but rapidly expanded in the 1930s. Originated by the entrepreneurs Captain Harry Warner, Fred Pontin and Billy Butlin, they coincided with the first statutory provision of holiday pay for all workers. Alongside the established 'formal' seaside leisure economy of the old resorts and holiday camps, there was an independent low-budget colonisation of under-valued coastal land from the late 19th century onwards. Shanty towns grew up from temporary holiday 'villages' along large parts of the Sussex coast, and elsewhere. Plotlands were a particular form of coastal residential development, arising from 20th-century

agricultural depression, low land values, and lack of planning controls. Typically, low-grade farmland was acquired by speculators, divided into rectilinear plots with a basic street system and then sold on to prospective occupants, who were then free to build as they chose on the individual plots (Rowley 2006). Perhaps the most whimsical example of plotland planning is at Jaywick, Essex, where the Brooklands Estate (named after the racing circuit in Surrey) was constructed by F C Stedman in the 1930s, seawards of the sea wall, with a ground plan based on the radiator grille of a Bentley. In 1953, the folly of developing seawards of the existing sea defences was shown all too plainly, for 35 people perished during the storm-surge of that year. It has defied initiatives for regeneration, but shows a unique and quirky 20th-century popular set of architectural and decorative styles, uninfluenced by high culture (Fig 4.58).

After World War II, the expansion of coastal caravan sites continued, for they evaded building bye-law restrictions: static caravans are in principle temporary and portable structures; but some caravans are in fact occupied year-round, in some cases by old or impoverished owners. The nominal seasonal occupation means that, at present, caravan sites do not figure in coastal management schemes.

Figure 4.58
The Brooklands Estate,
Jaywick, Essex. This house
could not fail to bring joy
into anyone's heart.

Coast-specific sites: hulks and wrecks

An intertidal hulk is the remains of a vessel abandoned at the end of its useful life, usually after removal of anything recyclable, while a wreck represents what survives from a vessel lost during active service. There is, however, little consistency in the distinction between hulks and wrecks in Local Authority Historic Environment Records. The vessels that became hulks and wrecks were, of course, of fundamental importance for trade, fishing or naval activities.

Some shallow-water or intertidal remains of vessels are early in date, but the majority of surviving examples date from the 18th century onwards. To give examples of prehistoric vessels: Bronze Age dug-outs were recorded from Preston Docks (NW1, 105), and a 2.08m long paddle or steering oar, dated 1255–998 cal BC, was found stratified in salt-marsh clays of the River Crouch at Canewdon, Essex (Wilkinson and Murphy 1995, 152–7). It was associated with plant macrofossils derived from salt-marsh and mudflats and presumably represents an overboard loss that was eventually stranded on a salt-marsh edge. Early medieval logboats from Warrington on the Mersey include one of oak, dated to cal AD 1090, and a second of elm, dated to cal AD 1000 (NW1, 75–6). A small dug-out boat from off the coast of Covehithe is dated to the Middle Saxon period, cal AD 775–892. Boat fragments from Easton Bavents and Buss Creek, Southwold, are late Saxon or medieval (S3, 13 and 38).

A few hulks or wrecks are of high status as Royal or Naval ships, though the majority are of trading vessels. The wreck of the *Grace Dieu* in the Hamble (destroyed 1439) is visible in Channel Coast Observatory photographs of 2008 (SE2, 50–1). Wrecks in the Hamble and Chichester Harbour include the *Royal George*, stranded 1757 on Chichester Flats (SE2, 52–3). More typical remains of several wooden ships lie on Mablethorpe beach, including the *Acorn*, a barque constructed in 1855 for carrying ice from Scandinavia to Grimsby for the fishing industry (YL3, 41).

Davies (2011) defines 199 hulk assemblages comprising 2 to more than 80 vessels: 9 with more than 20 vessels. A few hulks, as at Purton in the Severn Estuary, served a final function as components of sea defences: the hulks at Purton and Sharpness were emplaced during the 20th century to protect the Sharpness Ship Canal and Sharpness Docks from erosion (S1, 31). An Aerial Photographic regression study was undertaken to reconstruct the process of the accumulation and loss of hulks between 1945 and 2009 (S3). Most hulks were first beached at Purton after 1909, though the *Sally* (1780) was sunk to form a breakwater there in 1875 (S3, 5). By 1969 most hulks had been buried beneath canal dredgings, though new vessels, including Ferrous Concrete Barges had been emplaced. Of 45 hulks known to have been beached here, only 16 were still visible and largely intact in 1996. The Sharpness hulks were put in position in 1971–73, forming a regular linear barrier. The hulk assemblages include barges, hopper barges and lighters. Threats to both groups include erosion and vandalism. Hulks were also used to reinforce the riverbank at Lydney in the 1950s and 1960s. Eight vessels, probably including the Stroudwater barge *Nibley* were recorded, some of them filled with concrete (S4, 53). In the North-West, Mersey flats (shallow draft barges) occur with, at Sutton Locks, Cheshire, the flat *Daresbury*, built in 1772 for the Weaver Navigation company. A group of hulks, probably landing craft or barges, was incorporated into the embankment for the West Lynn Drain in Norfolk (N2, 227). At Forton Lake, Hampshire, the assemblage includes a motor fishing vessel, barges, pinnaces, ferries, motor gunboats, landing craft, lifeboats, as well as a minesweeper and bomb scow, both dating to World War II.

On the north bank of the River Coquet at Amble a group of hulks was recorded from aerial survey (NE4, fig 9) comprising six large vessels around 20 × 10m and two smaller ones 8 × 4m. These have variously been interpreted as herring boats or wherries used as coal lighters (NE1, 191). The largest hulk is carvel-, not clinker-built, and could perhaps be earlier than the others in the group (NE5, 183). They may have been abandoned when the coal port at Amble and Warkworth became defunct, though a longer period of abandonment is possible. The smaller vessels in particular appear to be possibly later in date and are probably fishing vessels.

In general wrecks and hulks relate to the pattern of trade then predominant, though also including fishing and naval vessels. In eastern England and Kent hulks of spritsail barges and wherries, which transported agricultural products from the region returning with cargoes of muck from London, for use as agricultural fertiliser, are common. Most wrecks from Breydon Water in Norfolk, usually on the edges of river channels, are thought to be wherries, keels and barges. Some wrecks are of World War II date, including HMS *Dungeness*, run aground off Happisburgh in 1940. At Hunstanton cliffs, the wreck of the *Sheraton*, a trawler built *c* 1910 and finally lost, after war service, in 1947 still survives on the beach, while the ice-ship *Vicuna* is visible at low tide at Holme-next-the-Sea (N1, 14). Isolated wooden fragments, and occasionally still coherent ship elements, presumed to have come from wrecks, were recorded at various locations during ground survey (N1, 78–9). Field survey has defined hulks in the Suffolk estuaries. They include the spritsail barge *Dover Castle* (1872), steam dredger *Holman Sutcliffe* (1890s), the barge *Three Sisters*, abandoned *c* 1932 in the Deben, and the *Tuesday of Rochester* in the Alde (S1, 18–22) (Fig 4.59). Ten wrecks around Foulness included a motorised barge and the trawler the *Florence* (E1, 15). Hulks on Mersea Island included small dinghies and skiffs, tenders for larger vessels, two lighters and two barges, including the *Victa* (1874) and probably the *Unity* (E1, 19; E5, 6).

In North Kent hulks dating back to the 18th century, or earlier, are concentrated within the Medway and Swale (NK1, 45) (Milne *et al*

1998). The schooner *Hans Egede* at Cliffe was the only non-barge sailing hulk located during survey (NK1, 62–3). Most hulks of commercial vessels were stem headed sailing barges, a type introduced after about 1840 (NK3, 21). They include the *Sirdar*, *Bessie Hart* and *Vera* in Milton Creek (NK5, 44). The paddle steamer *Medway Queen* of 1924 lies in Damhead Creek after serving as a minesweeper in World War II. The remains of two mini-submarines are at the same location, laid up in 1920; bulkheads are visible (NK6, 69).

In Cornwall, a wreck believed to be of the barque *Antoinette*, lost on the Doon Bar in the Camel Estuary in 1895, bound from Newport for Santos with a load of coal, was recorded before destruction as a hazard to navigation in 2010 (ADS Collection 1050; doi: 10.5284/1000403). On the Fowey Estuary there are records of wrecks and hulks, with a concentration of 19th–20th century hulks at Mixtow and Pont Pills, probably including the schooner *Jane Slade* of 1870. The sailing barge *Sunbeam* of 1913 was abandoned in Porth Navas Creek around 1937 and remains survive (Reynolds 2000, 26).

Offshore wrecks were generally beyond the scope of the RCZAS, but known wrecks in the Solent were investigated by divers as part of the New Forest survey (NF1, 29–39). They include

Figure 4.59
Alde Estuary, Suffolk. Hulk of the sailing barge Tuesday of Rochester.

HMS *Assurance* (1753) off the Needles, and the *Pommone* (1805). World War II German and British aircraft wrecks are also recorded, including the motor dredger *Margaret Smith* (1943, wrecked 1978), the tramp steamer SS *Serrana* (1905, torpedoed 1918), the Dutch two-masted schooner *Fenna* (1863), the steam barge SS *Ceres* (1875) and the armed merchantman SS *War Knight* (1917), which sank in 1918 after a collision (NF2, 68–71).

Trackways and causeways

Ships and boats provided the main means of coastal communication, but old terrestrial routes are also detectable. Yates (2007) and Bell (2013) have identified coaxial field systems of Bronze Age date with associated droveways leading to coastal wetlands, for example at Flag Fen, Cambridgeshire, and around the Severn Estuary. In Norfolk derelict wooden bridges crossing creeks on salt-marshes, and trackways composed of recent building rubble are recorded from numerous locations, for example at Warham, including 'Cocklestrand Drove' (N1, 77). Although plainly recent in their present form, these structures perpetuate older routeways, largely related to grazing on the marsh. On Breydon Water a timber trackway and revetment was recorded. It comprised two post rows, aligned NW–SE, with wattling between the rows, seemingly truncated by a breakwater or revetment. The structure has a prehistoric appearance, but in this context must be Saxon or perhaps much later (N1, 111). Between Snape or Friston and Iken, in Suffolk, a linear spread of gravel marks a causeway, shown in roughly this position in the 1st edition OS map (S1, 13). At Snape Warren horizontal unworked timbers marked the line of a causeway, shown on the 1st edition OS map: local information suggested a 19th-century origin, to permit carts to cross the marsh (S2, 3).

In Essex, river crossings constructed of timber and gravel, as at Fambridge and Hullbridge on the Crouch (Fig 4.60), represent a major effort of construction. Plainly fords, not bridges, they are at least of medieval origin, possibly earlier, but have not been archaeologically investigated (Wilkinson and Murphy 1995, 201–7).

The Broomway, a trackway around a quarter of a mile offshore from the Foulness archipelago, provided a wagon route between the islands until construction of a road bridge in 1922. It was used to transport goods, including fresh water, which had to be imported. Its origin is unknown but it could perpetuate a terrestrial route pre-dating medieval sea-level rise (E1, 4).

Figure 4.60
Hullbridge, Essex.
The Hullbridge itself is visible as a large timber alignment, not a bridge but a substantially constructed ford.

Intertidal trackways were recorded on the southern edge of Havengore Island: a track of brick rubble overlay an earlier (undated) wattlework track. A similar wattlework track connected Wakering Stairs to the Broomway (E1, 16). The Mersea Island Strood Causeway is scientifically dated to the Middle Saxon period (E1, 6). A similar structure, known as the Wadeway, connected Hayling Island with the mainland of Hampshire. It is now incomplete, but originally comprised a gravel and timber causeway, very probably kept going, over centuries or even millennia, to maintain a route. Its origin is of unknown date, but it is on the line of what must have been a route onto what was the Hayling peninsula before relative sea-level rise.

The history of ferries is likewise poorly understood. In North Kent there are historic crossings and a causeway including Harty Ferry (originating pre-18th century) and a causeway, from at least the 17th century, crossing Higham Common. The latter was reputedly Claudius' crossing point for the Thames (NK1, 63). The Old Passage Ferry in the Severn was replaced by the original Severn bridge in 1966 (S1, 23).

Terrestrial coastal routes are as poorly understood archaeologically as are minor landings (see above). The superficially modern appearance of many, combined with the logistical problems of excavation, has discouraged investigation. We have no idea how ancient they may be; but we can be confident that they were as important economically as inland routes.

5

Research priorities

Research priorities change through time. Fulford *et al* (1997, 215–32) present a programme of priorities that seemed appropriate in the late 1990s. This chapter will refer to them – many have now been achieved – but it will also take account of the English Heritage Marine and Maritime Research Framework (Ransley and Sturt 2013), Regional Archaeological Research Frameworks (*see* Appendix) and the RCZAS reports themselves.

Fulford and his colleagues say that 'in the absence of entire categories of coastal monuments [from the then-NMR (now NRHE), and HERs] such as flood and sea defences, or seaside and harbour architecture, and of coastal industries such as fishing, ship-building, and leisure, there is insufficient data with which to begin to characterise England's coastal heritage'. Even in 1997 this stricture was excessively harsh, for plenty of information was available, but there were inadequacies. As outlined in Chapter 3, the RCZAS have helped to fill gaps: we now have extensive and reliable information on all these aspects of the past. Fulford *et al* also note the disparate character of earlier coastal surveys, which often focused on particular periods or categories of site. Partly due to a broader perception of what the historic environment comprises, the RCZAS attempted to avoid such biases, aiming to record everything equally well.

Period priorities

The Palaeolithic

Knowledge of lower Palaeolithic sites on coasts has increased very substantially since 1997, though plainly their current topographic situation often does not bear much relation to their original environs. The sites at Happisburgh and Pakefield in East Anglia, however, did lie within the coastal zone when occupied. They provide the earliest evidence for a pre-modern human presence in Western Europe (back to 0.78–0.99 million years: Parfitt *et al* 2005 and 2010). Such a long time period was not dreamt of in 1997. Moreover, since then, systematic archaeological investigation of offshore Palaeolithic contexts has begun, notably at Area 240, off Great Yarmouth, Norfolk (Wessex Archaeology 2011). Neither of these initial developments followed from the RCZAS, but from the British Museum's Ancient Human Occupation of Britain (AHOB) project and from the MALSF. However, there is a need to link together stratigraphic data from infilled former river channels inland, those on the shore and those in the shallow marine zone (to 1km offshore) to create a unified reconstruction, to assess the ways in which coastlines may have been exploited in the Palaeolithic, and to evaluate the vulnerability of sites to erosion. This will be supported as part of EH's National Heritage Protection Plan (Project 3A3.201). The intertidal/shallow-water sub-tidal so-called 'white zone' (where data acquisition is difficult) presents the most problems. However, a combination of laser scanning, small draught multibeam sonar, swath bathymetry, sub-bottom profiling, magnetometry and perhaps conductivity will be developed, based on a small shallow-draught vessel platform. As in the case of offshore work, innovative methods will need developing.

An improved reporting network for unstratified beach finds – sometimes lithics but more often faunal remains showing evidence of butchery – will also be established. Beach finds should not be dismissed as out of context and so archaeologically irrelevant. On the contrary, they can indicate the proximity of eroding coastal sites, for all periods, but especially for this one. The Portable Antiquities Scheme has resulted in the reporting of some coastal finds, but there is a need to develop better reporting mechanisms for offshore finds made by fishermen and for coastal finds made by beach-walkers, comparable to that developed for offshore aggregate extraction (Wessex Archaeology 2005b). A Fishing Industry Finds Reporting Protocol is being trialled regionally for the offshore area of the county of Sussex, in collaboration with the Sussex Inshore Fisheries and Conservation Authority (IFCA), funded by EH.

Although we now have a better understanding of upper Palaeolithic palaeogeography and, from offshore contexts, information on fauna and fauna (Peeters *et al* 2009, 21–4), new sites of this period have proved hard to find in the intertidal zone. In fact no artefacts of this period are as yet known from *offshore* in the North Sea. This apparent upper Palaeolithic hiatus is hard to explain, for individual lithics are known from several coastal locations. Most probably, the smaller size of lithics of this period has meant that they are far less likely to be captured in fishing nets, or seen and collected on beaches, than lower Palaeolithic artefacts. Terrestrial cave sites have produced the largest assemblages of material of this period, suggesting one possible line of investigation. In areas of limestone terrain, for example around Lundy, marine geophysical survey indicates a submerged eroded karstic surface which is likely to include caves. These would have been habitable during

sea-level low stands. Archaeological investigation of submerged caves will present methodological and logistical difficulties, but they are likely to produce primary contexts undisturbed by 19th-century archaeologists, as they have been on land.

The Mesolithic and later prehistory: the submerged forests and peat.

Fulford *et al* (1997, 216–17) recommend development of a national base-line survey of peats and submerged forests. This has now been achieved (Hazell 2008) and some of these deposits have been dated by radiocarbon as part of the RCZAS. In 1997 it seemed that these peats would provide information on the hunter-gatherer/agricultural transition and the presence of sedentary late Mesolithic communities. The definition of huts at Howick Burn certainly establishes the latter (Waddington 2007), while the submerged Mesolithic site at Bouldnor Cliff could likewise represent more permanent settlement (Momber *et al* 2011). However, the intertidal peats still require more detailed investigation focusing on that part of the sequence dating to around the Mesolithic/Neolithic transition. Study of submerged forests was initiated by Clement Reid in the late 19th century, who was the first to appreciate their significance in terms of defining the extent of offshore submerged landscapes (Figs 5.1–5.3). He went on to

Figure 5.1
Hightown, Lancashire.
Submerged prehistoric
forest.

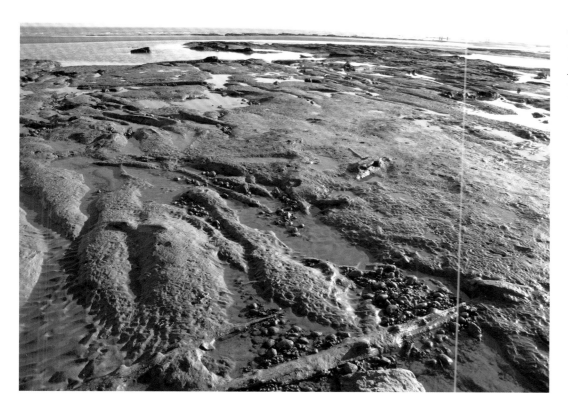

Figure 5.2
Bulverhythe, Hastings.
Submerged prehistoric
forests occur all round the
country, but are of widely
differing dates.

produce the first tentative map of the Mesolithic landscape of the North Sea (Reid 1913). Modern offshore survey and investigation have provided increasingly well-defined models of Mesolithic landscapes, helping to place the coastal and terrestrial record in context, notably by Gaffney *et al* (2007).

Other significant finds include human and animal footprints on intertidal sediment shelves (at Low Hauxley, Oldbury-on-Severn, Formby and Crosby), reflecting hunting and gathering activities. Currently EH's Remote Sensing Team is proposing to trial multi-image photo-grammetry to record the exposed footprints and their surrounding context to around 1–2mm point resolution. Since these footprints represent an unusual opportunity to see 'fossil' human activity – food collection, hunting, travelling and destinations – recording to the best possible standard is necessary.

We can see mobility directly from these footprints, but transportation would also have involved the use of boats. A few logboats of probable Mesolithic date are known from England, but dating is often uncertain and, besides, more sophisticated and stable craft such as skin boats must have been in use for longer crossings, for example to the Isles of

PLATE VIII

HORACE BOLINGBROKE WOODWARD

GEORGE WILLIAM LAMPLUGH

EDWIN TULLY NEWTON

CLEMENT REID

Photo Elliott & Fry

Figure 5.3
Clement Reid (bottom right)
and colleagues from the
British Geological Survey,
from Flett (1937). H B
Woodward (top left) was
among his collaborators
in survey in the East
of England.

Scilly. Martin Bell and Graeme Warren (2013) emphasise the need for more information on Mesolithic seafaring. They also highlight factors that influenced Mesolithic settlement patterns affecting the preservation and visibility of sites as an area requiring study. In addition, more data are required on the exploitation of marine and terrestrial plant and animal resources on coasts. The Mesolithic is often characterised as being focused on coastal resources, but the evidence for this is sparse in lowland England. Detailed investigation and extensive programmes of sieving of deposits at habitation sites is needed, especially where waterlogged layers occur. Finally, comparison is needed between uses of food resources in the Mesolithic and the Neolithic in order to gain better understanding of the transition between the two.

Later prehistory

Results from the intertidal Neolithic site of The Stumble in the Blackwater Estuary, Essex, have been outlined above. Though it was within the coastal zone when occupied, there is no evidence to suggest that its residents exploited marine resources. It was a valley floor site with a farming, hunting and gathering economy, which just happened to be submerged later. Its significance lies in the fact that submergence resulted in a near-constant depositional water-logged environment, in which charred plant material and ceramics did not degrade signi-ficantly owing to physical and biotic processes. It produced vastly more material than nearby contemporary sites on the river terraces, which have remained terrestrial for at least 6000 years (Wilkinson and Murphy 1995; Wilkinson et al 2012). During excavation in the 1980s it was thought that sites of this type might be widespread but, in fact, they are scarcely known outside the Essex and the North Kent coasts. Consequently sites of this type are of much greater significance than was realised at first. As noted in Chapter 3, monitoring of the site continued but has now ceased. Sites of this type provide unusual opportunities for investigation of the Neolithic, and so further monitoring and investigation is needed.

The Bronze Age on the coast, as elsewhere, is dominated by monumentality. The cist and cremation cemetery at Low Hauxley, Northumberland (NE3), and the timber circle at Holme-next-the-Sea, Norfolk (Brennand and

Taylor 2003), are noted in Chapter 4. Both sites are eroding and (in the case of Holme) degrading, due to bioactivity, and in both cases further monitoring and excavation are required. Cliff erosion is likely to result in further loss of barrows and related structures and it is hard to see how recording can be achieved other than by the voluntary sector (see Chapter 6).

Settlement sites are likewise noted above. Excavations have been undertaken at Brean Down and Gwithian. Brean Down is protected by coastal defences emplaced by English Heritage and the National Trust: periodic monitoring indicates that this has maintained the site's stability over the last 26 years. The main area of Bronze Age and early medieval sites at Gwithian are 700m from the sea, but eroding cliff and dune systems have produced palaeoenvironmental data and a possible prehistoric field wall (M Bell, pers comm). More generally, Bell and Brown (2009) have defined numerous archaeological sites exposed by dune erosion in southern England. On lowland estuarine coasts further evidence for ritual structures, timber alignments and boats is likely to be obtained, but persistent monitoring is needed to ensure this. Inter-ventions undertaken at development as part of the planning system are likely to reveal other structures, such as the timber platform at Shinewater Park, Sussex. There is evidence for salt production beginning in the middle Bronze Age at Brean Down, but further survey and radiocarbon dating is needed to differentiate early sites from the far more numerous Iron Age and Roman salterns. Despite concentrations of Neolithic and Bronze Age finds and sites in the coastal zone the evidence for exploitation of marine food resources is sparse and confined to a few sites (Sturt and Van de Noort 2013). Further investigation of sites where soil conditions would permit survival of fish-bone and shell is needed. This also applies to the Iron Age, during which cultural prohibition of such foods has been suggested. Is the apparent scarcity of the remains of shellfish, seabirds, and mammals on English sites real and, if so, why would people choose not to make use of these plentiful resources (Hill and Willis 2013, 83)?

Britain was an archipelago by the Neolithic and offshore islands such as Scilly were occupied, at least seasonally. Sea-going boats must have existed. Although logboats and sewn-plank boats are relatively well recorded from the early Bronze Age (after around 2000 cal BC), remains of Neolithic boats are unknown. This may have related to their form of construction (perhaps hide or skin-covered wooden frames), which would not survive well. A priority is careful investigation of sites where wood preservation could be expected (Sturt and Van de Noort 2013). Similarly, for the Iron Age our ignorance of vessels is profound. Is it possible or efficient for researchers to target areas of high potential to increase finding boat remains, wrecks or lost cargoes [of Iron Age date] (Hill and Willis 2013, 87)?

Aerial photography in the coastal zone has defined extensive field and enclosure systems, especially in the south and east, and many of them originated in later prehistory. In some cases they do not relate obviously to the coast, but might represent the seawards edge of a more extensive planned landscape. However, Bell (2013) provides a model linking Bronze Age activity on the 'upland' with seasonal grazing of coastal wetlands around the Severn Estuary. Moreover, excavation shows that some settlements, such as the 2nd–1st century BC example at Hallen, can also be interpreted as seasonally occupied sites based on grazing. Other features include forts – both promontory forts on cliffs and in low-lying wetlands – burial sites, manufacturing sites and trading settlements. The 'proto-ports' at Hengistbury Head, Mount Batten, Poole Harbour and Meols have been investigated in some detail since 1997. However, assemblages of coins from other locations hint at the existence of other trading sites. Long-term monitoring is needed at these so-called 'productive' sites. It appears that much of the evidence for salt production on the Lincolnshire coast has been lost to erosion, but elsewhere salterns are known from many locations including Essex, North Kent and Poole Harbour. Continuing erosion will undoubtedly reveal new sites, but the principal need is for large-scale excavation to clarify the means of production and chronology.

Roman

The enthusiasm of Romans, and the people they influenced culturally, for seafood and fish is well known (Murphy 2009, 84): shells and fish-bones occur in profusion at excavations. Likewise, imported artefacts point to large-scale overseas trade. Beyond that they were also a maritime military people. Yet, despite

almost four centuries of seafaring, few wrecks and port facilities are known from England, and these are mostly from London and the Thames. Apart from military defences and salterns, the Romans are strangely invisible on the English coast, despite the durability and conspicuous character of their material culture. Undoubtedly some sites have been lost to erosion, but further work to locate and characterise Roman port sites, at the Saxon Shore forts and elsewhere, is needed. Finding Roman period wrecks is likewise problematic. It seems probable that offshore timber wrecks have not survived after two millennia of seabed change around England (unlike in the more tranquil Mediterranean) but their durable cargoes, including amphorae or the Samian ware (as, for example, from Pudding Pan Sands off North Kent), should be obvious. The odd thing is that there are so few known sites. Actual ship remains are more likely to be found in sediments infilling embayments, estuaries and ports. Religious and ritual practice on coasts at this time also needs more study (Walsh 2013).

Roman period salt production is much better defined, with sites known from most parts of the country. However, they have generally not been investigated extensively. Excavation of a 44-hectare site at Stanford-le-Hope Wharf, Essex, provided a detailed picture of production between the middle Iron Age and 3rd centuries, but this is uncommon (see above). This excavation was undertaken in advance of ground-level reduction, sea wall construction and breaching, as ecological mitigation for the London Gateway Port Development (Fig 5.4). It seems probable that other Managed Realignment Schemes will provide opportunities for excavation and recording. Reclamation of coastal wetlands appears to have been focused around the Severn Estuary, leaving many areas unclaimed and used for salt production and grazing. Future excavations may add to the known areas of Roman reclamation elsewhere.

Post-Roman: Saxon and medieval

Fulford et al (1997, 218–19) review these periods very briefly, insisting on a general lack of archaeological data. This is surprising, given the extensive structural evidence around the country. The RCZAS and other projects have provided information on stationary fishing structures, reclamation, and religious establishments, all distinctive Anglo-Saxon features of

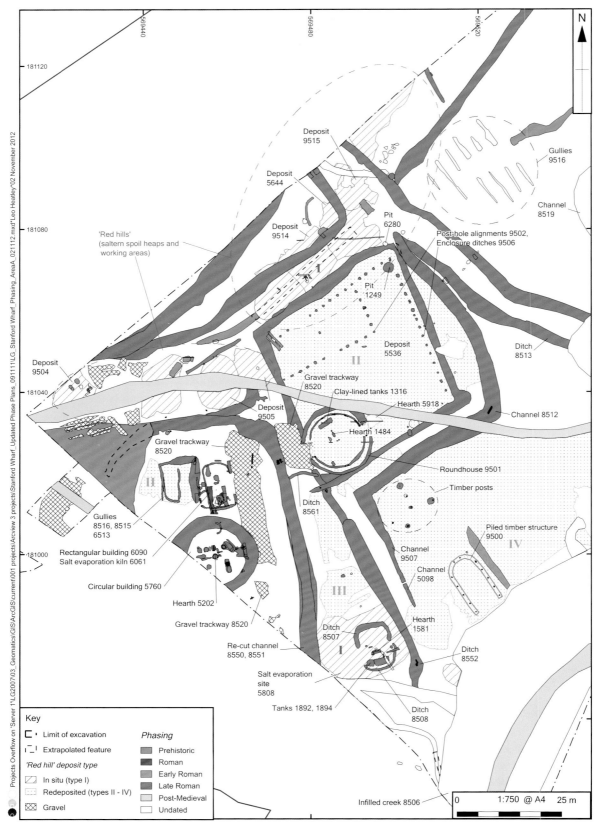

Stanford Wharf - Plan of Area A

the English coast. Timber fish traps of various types have been dated by radiocarbon on the east coast and in the Severn. In both areas most dates cluster in the 7th–13th centuries, mainly in the earlier part of this period, with some dates from the Severn in the post-medieval period. Fish traps are often associated with nearby monastic foundations and so could be considered the precursors of monastic fish ponds. This expansion of stationary fishing seems, on the present evidence, to be part of the 'long 8th century', a period of economic innovation across much of Europe, but more dates are needed to assess whether this pattern applies elsewhere in England.

Documentary sources show that renewed land-claim began in the middle to late Saxon period, when parts of the Somerset coast, Fenland, North Kent and Romney Marsh were embanked. From the 12th century onwards, the general trend was towards reclamation in back-fens and coastal marshes for mixed agriculture, though with local and regional specialisation in production (Rippon 2000). Archaeological evidence for the dates of banks is, however, slight. One sea bank at Foulness is dated by dendrochronology to the 15th century. A sequence (but not a chronology) of embankment can be inferred from the forms and pattern of systems of embankment seen in aerial photographs. Further scientific dating of banks is needed.

Monastic sites were frequently located on coasts and hence several are now at risk of erosion. They include St Cuthbert's Island at Lindisfarne, Whitby Abbey and Cockersand Abbey, Lancashire (Fig 5.5). Excavation of vulnerable buried sites is needed where funding can be found. Likewise some parochial churches will be at risk in the next 20–50 years, for example at Covehithe and Happisburgh in East Anglia (see Fig 4.19). The costs, practicalities and logistics of recording an eroding large standing medieval church, underlying deposits related to its predecessors, and the burials in its churchyard, are considerable. It is hard to see, at present, how this could be achieved.

Small havens and landings have been discussed above. Some were constructed for the use of specific industries and so can be dated. The vast majority, however, have produced no dating evidence: many could be of medieval or much earlier origin. A systematic programme of recording and dating is needed. Medieval salterns are recorded from many locations, though the chronology of these sites is often poor, so some could be later (Fig 5.6). More detailed investigation will be needed to clarify this, and investigations related to Managed Realignment provide the best potential opportunities.

Wics (trading settlements) were not considered specifically during the RCZAS, for they are often under modern ports, or else now

Figure 5.4 (facing page) Stanford-le-Hope Wharf, Essex. Phase plan of extensive excavation of a middle Iron Age to Roman saltern site, recorded as part of a habitat creation scheme. (Courtesy of DP World London Gateway and Oxford Archaeology)

Figure 5.5 Cockersand, Lancashire. Chapter House of S Maria de Marisco [of the marsh], a 12th-century priory (later, monastery) of Premonstratensian Canons, 12th century.

Figure 5.6
North Wootton, Norfolk.
A medieval sleeching mound
(in the background) and
palaeochannel, now lying
within grazing marsh.

well inland, owing to coastal change. Carver and Loveluck (2013) highlight them as research priorities. The medieval period also saw the development of more sophisticated, and larger, vessels capable of long-distance voyages, maritime warfare and large-scale offshore fishing (Carver and Loveluck 2013). Compared to earlier periods more wrecks have been recorded, but still very few. The criteria noted above for earlier periods need to be applied to find more.

Post-medieval

The review of this period by Mike Fulford and his colleagues is remarkably brief (Fulford *et al* 1997, 219). The very term 'Post-medieval' reflects how our perception of the historic environment has changed since 1997. Today we would refer to this period as 'Early Modern' and 'Modern'. Their term, using the word 'post', places these periods at the periphery of archaeological interest, whereas the present terminology brings it bang up to the present as part of a continuous process. A new definition of research priorities is required, most simply defined by bullet points:

- Sea defences, land-claim and erosion. As noted in Chapter 4 there is good information on reclamation and later losses of land, mainly from historical sources. However, in many areas the archaeological evidence for the chronology of reclamation is very limited. Managed Realignment Schemes result in damage or destruction to

sea defences, but curatorial archaeologists often do not have any factual basis on how to respond to such schemes. Are they ancient, or are they not? More data are required. In the near-shore zone, remains of eroded sites persist. As work at Dunwich has shown these are not necessarily just scattered rubble, but include structural remains that can now be investigated using marine geophysical and other techniques.

- Ports, quays and shipbuilding. The major ports are, in general, well recorded from both documentary and structural sources. However, settings on quays and piers (such as capstans, mooring posts and other features) that were integral to their function have often not been noticed nor included in designation descriptions and hence are vulnerable to future renovations. There is a risk that they might be regarded simply as decorative 'maritime' features that might be removed or relocated. They require further assessment with a view to specific designation. Minor ports, havens and landings are, as a rule, not understood at all, though they represent a pattern of trade that was in the past essential (*see* Chapter 4). Many have been abandoned and so, unlike at the major ports, substantial remains of early phases of waterfronts should survive. Targeted programmes of investigation to understand their roles, chronology and phases of development are required. Shipbuilding (and the related trades of ship-breaking and repair) were formerly far more widespread than today, especially before the advent of iron and steel ships. Some locations have been investigated but far more remain poorly understood.

- Fishing and fishing structures. The coastal evidence for fisheries includes structural evidence – fish traps, shiels, fish-smoking houses, fishing houses, huers' huts and fish cellars in different parts of the country – and archaeological deposits related to fish processing. Further work to characterise the regional variation in structures, and their adaptation to individual fisheries is needed. Many are minor and humble buildings whose significance could easily be overlooked. In some areas, for example in the Severn Estuary, there remain a few fishermen still using traditional trapping techniques, and this ethno-historical source of information should not be overlooked. Some fisheries have almost disappeared from the historical consciousness of the country, especially whaling, and need especial emphasis.

- Salt. Compared to the information available in 1997 several projects have provided information on salterns, notably in the New Forest. Elsewhere Early Modern salt production and its relationship to the wider economy require study.

- Extractive industries. Several surveys of sites producing alum, coal, limestone and jet have provided the data highlighted as being needed in 1997, especially in the North-East and Yorkshire. Elsewhere the physical remains of extraction of these minerals, and of building stone, have scarcely been investigated.

- Military defence. Military defences were specifically excluded from the 1997 report. However, the RCZAS have produced a vast amount of data, especially on defences of the 20th century. The highest priority at present is differentiation of World War I from World War II structures and thematic designation of the former. This would be especially apposite given the approaching centenary of World War I in 2014. For earlier periods the less monumental structures – principally earthwork defences – have received insufficient attention.

- Religious foundations. As noted above, the majority of Christian sites of these types relate to earlier periods. However, some late phases of construction that lie in locations vulnerable to erosion may be locally characteristic, and so require recording.

- Hulks and wrecks. In common with minor landings, hulks and wrecks are only now receiving proper academic attention. The NHPP project *Hulk Assemblages: Assessing the National Context* marks a step forward, but it should be noted that it focuses on assemblages, not on the numerous isolated hulks around the coast. These also require assessment.

- The development of a global economy. This is such a vast topic, which now underpins the economic life of the country, that it almost defies definition of research priorities (Dellino-Musgrave and Ransley

2013). Research approaches have tended to be specific, addressing particular aspects of the process, for example port and ship development. Perhaps the main need is for specialists in, for example, the development of warehouses at ports and offshore wrecks, to perceive the need to talk with specialists in other areas of maritime archaeology.

Thematic issues

Environmental potential

Even in 1997 (Fulford *et al* 1997, 219) the techniques of environmental archaeology were widely applied during archaeological investigations. They are now routine and standard at both onshore and offshore sites. The range of palaeoecological indicators used in archaeology has, however, expanded since 1997, to include microscopic proxy palaeoecological indicators, such as diatoms and foraminifers. Other key economic issues include investigating the farming basis of coastal settlements. For example, studies of Fenland coastal Anglo-Saxon sites show that they were not simply seasonal grazing settlements but included crop production – even before sea-banks were raised (Murphy 2005a; 2005b). Moreover this agricultural production was specifically adapted to the undefended coast by focusing on salt-tolerant crops such as barley. Further work of this sort will help clarify the earliest phases of coastal wetland colonisation.

Scientific dating

Dendrochronological evidence shows that the Bronze Age timber structure at Holme-next-the-Sea was built in the spring or early summer of 2049 BC (Brennand and Taylor 2003). This result alone fulfils Fulford *et al*'s assertion that 'dendrochronology … will help to evaluate … change on timescales which are more relevant to issues of human … perception and knowledge than chronologies derived from radiocarbon dating …'. However, such precision is not possible for most intertidal wooden structures since they either were not constructed of oak, or were of young oak stems including insufficient rings for dating. Given that numerous submerged forests of late Mesolithic date survive, a programme of dendrochronological analysis would help to improve chronology for this period. Radiocarbon dating, however, remains a key chronological method for the Holocene. It defined, for example the initiation of fish-trap construction in the Middle Saxon period (Chadwick and Catchpole 2010; Murphy 2010).

For the earliest periods, especially the Palaeolithic, other new dating techniques require emphasis, notably Optically Stimulated Luminescence (OSL). It is applicable to any sediment unit including quartz grains and provides dates for units pre-dating the applicability of radiocarbon. The results are plainly not very precise, with wide error ranges, but do place sediments within Marine Isotope Stages that are undatable by other means. For example, sediments dating back to $283 \pm 56 \times 10^3$ years were obtained from the offshore site of Area 240, off Great Yarmouth (Toms 2011).

Coastline change

Section 9.3.3 in Fulford *et al* (1997) focuses very specifically on the Roman period and on changes experienced during the late 13th–16th centuries. As noted in Chapter 3, we now have data extending back to almost 1 million years. It is probable that further understanding of coastal change in earlier prehistory will come principally from offshore investigations. For the Palaeolithic and Mesolithic a suite of techniques (seismic profiling, sediment coring, palaeoecological analysis, radiocarbon and luminescence dating) is available and needs wider application to submerged sites on the continental shelf landscape (Westley and Bailey 2013; Bell and Warren 2013). These techniques permit large-scale landscape reconstructions but a key issue is to 'populate' offshore landscapes and to interpret change in ways that would have affected human behaviour: development of models of preferred settlement location is needed. A key question for the Neolithic colonisation of England is when did the last islands in the North Sea become submerged? (Sturt and Van de Noort 2013). By the Roman period questions about coastal change become more specialised and local. It is plain that there has been significant change since then. Reconstruction of the coast of the Roman period, the sites on it and, indeed, ports of the period will require more specific regional investigations which cumulatively will produce an overall model (Walsh 2013).

The late 13th to 16th centuries were described by Fulford *et al* (1997) as 'the critical centuries in the evolution of our present coastline'. Reclaimed land was lost, not to be recovered until later, and entire ports were either physically destroyed, as at Dunwich, or silted up to become unusable. This is plainly of archaeological and historical significance, but in addition the experience of people at that time has a direct resonance for the 21st century, when comparable changes may be expected once more. Defra (2010) included a historical section provided by EH, discussing the loss of Dunwich in its 'Adapting to Coastal Change' policy framework. Reflecting on past experience, and understanding that coasts have always been dynamic, helps to make people better prepared for future change (*see* Chapter 6).

The habitability of coastal wetlands was locally variable, so that adjacent wetlands would have provided very different conditions at the same time. In Sussex, for example, the coast includes several estuaries and embayments. At some periods (though not synchronously) they were isolated from the sea by shingle spits and barriers, which formed due to long-shore drift of sediment in an easterly direction from around Selsey Bill. This diverted the mouths of rivers eastwards, thus allowing mudflats, salt-marsh and, ultimately, freshwater wetlands to form in back-barrier environments. As the peat surface dried out, fen-carr woodland developed and the areas were inhabited. After the late Bronze Age, the peat surfaces were overlain by marine and estuarine sediments, though again not synchronously. This probably reflects breaching of shingle barriers at different times at separate places, owing to local factors, in later prehistory (Woodcock 2003). Detailed investigation of coastlines will be required to elucidate similar local variability, and human responses to it, elsewhere.

Coastal and flood defence

As noted above, there have been considerable advances in our understanding of the chronology of land-claim and, indeed, of land loss. Some Roman sea defences are known from the Severn, later Anglo-Saxon and medieval defences are known to have existed (mainly from documentary sources) in many places, and the later history of land-claim can be inferred by matching extant structures with historical records. Losses of claimed land, mainly from the 14th century onwards, have likewise been inferred. The priority now is detailed study of specific areas, such as that undertaken on Romney Marsh, to provide a contextual interpretation of surviving earthworks. The sea walls of England are, taken together, the most extensive archaeological earthworks in the country yet they remain poorly understood.

Coastal settlement and the maritime cultural landscape

For the Mesolithic and Bronze Age Bell (2007; 2013) has proposed models of coastal land use, based on data from inland and coastal sites. From later prehistory onwards the RCZAS have provided, on the one hand, extensive information on patterns of field systems and settlements within the coastal zone (mainly from aerial photographic evidence) and, on the other, shoreline-based activities such as fish traps and salterns. Linking the two together to develop models for a maritime cultural landscape has scarcely been attempted. Similarly, the lack or paucity of evidence for exploitation of marine resources in later prehistory requires further consideration. Is it

real? Could it relate to the use of the sea for ceremonial or burial purposes? Overall, the need is for more developed interpretation of the data we now have.

Early ports, quays and wharves

As noted above, coastal landings were of fundamental importance for trade in the past but the chronology of such sites is very poorly understood (Fig 5.7). Formal archaeological excavation of a selection of suspected early sites is needed. Some may have prehistoric origins, and a long history of modification, but we really do not know. Opportunities to undertake this work as part of archaeological programmes alongside Flood and Coastal Erosion Management schemes must be exploited.

The fishing industry

Compared to 1997 we now have extensive information on the distribution, character, and chronology of stationary fish traps. The overall impression is that there was a major phase of construction during the 'long 8th century', often in association with monastic sites, and part of a wider economic expansion. Subsequently, construction and use continued

Figure 5.7
Craster, Northumberland.
A small rock-cut harbour.

through the medieval period and onwards, though Early Modern structures are surprisingly rare. Further radiocarbon dating is needed. Most of the evidence for marine fisheries has come from fish bones retrieved from archaeological excavations, but little attention has been paid to fish-processing facilities.

Salt-making

Again, recent research has expanded the number of known sites, from the later Bronze Age onwards, while excavations such as that at Stanford-le-Hope, Essex, have provided detailed information on changes in production over time. Further large-scale excavation is needed. Medieval and Early Modern salterns are known from both documentary and field evidence, but there has been little archaeological investigation.

Quarrying

Extensive surveys of Early Modern extraction for ironstone, alum, building stone and other minerals have been completed during the RCZAS and other studies. However, Roman (and occasionally earlier) coastal extraction sites for jet, Kimmeridge shale and building stones remain largely uninvestigated. In many cases evidence for the earliest extraction will have been obliterated by later quarrying, but it should be sought.

Ships, boats and craft

Systematic national assessment of Early Modern assemblages of hulks and wrecks has now begun (Davies 2011). This work needs completion to obtain a full national record and extension to record isolated remains of vessels. Subsequently, the results require prioritisation to identify sites needing detailed survey and, very occasionally, lifting and conservation. Ship remains in harbours that are being developed for recreational use are especially vulnerable to a desire to 'tidy things up' and it is therefore a high priority to identify and conserve the most significant examples. Plainly *any* wreck or hulk of earlier date requires detailed study and recording.

In addition, finds representing material derived by erosion from offshore wrecks (especially substantial timber elements) are frequently found on beaches. Though strictly speaking unstratified, these finds can be informative. In some cases they are incorporated into Historic Environment Records, but this is not universal. There is no clear definition of roles and responsibilities in relation to recording such finds, and little museum or archive capacity. 'Emergency' funding for conservation is lacking. These problems relate to most maritime finds, but coastal finds also require consideration.

Shipbuilding

Sites related to the construction, maintenance and breaking of ships have been noted in Chapter 3. Given the economic significance of these activities, at least from the Bronze Age onwards if not earlier, archaeological investigation of such sites has been limited, largely related to excavations associated with development control.

Regional priorities

Fulford *et al* (1997, 231–3) set out a short list of regional priorities for further survey and research. They emphasise that their list is 'no more than indicative, serving to stimulate debate at a regional level'. Since then there *has* been debate at that level, with publication of Regional Archaeological Research Frameworks (RARF).

Subsequent work has, to some extent, filled the gaps in information they identify. For example, information on submerged peats and forests has been presented at a national level by Hazell (2008 and the associated database), and several of the RCZAS surveys have produced new radiocarbon dates for these sites. The basic data have now been assembled. What is needed now is an appraisal of which sites are of significance for investigating well-defined research questions, such as: 'How did the landscape change around the Mesolithic-Neolithic transition?' Sites that have associated archaeology or are close to Neolithic and Mesolithic sites have the highest potential for producing relevant palaeoenvironmental data.

In other cases, as a result of the RCZAS, it is possible to refine their generalisations (for example 'Coastal dunes: monitor Northumberland coast') to studies of specific sites or known sites of special significance, such as Low Hauxley. In most cases this moves beyond survey to more specific investigation. Several of the reports conclude with recommendations for the establishment of volunteer groups to carry forward long-term monitoring, so this should be seen as a national priority, which will not be reiterated here (*see also* Chapter 6).

Regional research priorities are summarised in the Appendix. Approaches to assessing risk and managing the coastal historic environment at a national level are considered in the following chapter.

6

Managing England's coastal heritage

As William Morris remarked, 'The past is not dead, it lives within us, and will be alive in the future we are helping to create'. Managing and conserving the historic environment contributes to making a future. As previous chapters have shown, we have many wonderful, valuable and precious things from the past around our coasts. They contribute to our sense of place, to national identity and to the meaning of living on an island. They are joys that enhance life. We must hand them on. Yet the coast was never static and future climate change will mean that many historic assets are increasingly at risk. We habitually refer to the 'Sea Front' as we might have done in 1914–18 to the 'Western Front' – the place where familiar territory meets the foe; but it might be a misapprehension to see the sea simply as an enemy.

First, we have to accept the inevitable consequences of coastal change. Coasts are dynamic, but most historic assets are fixed, which inevitably means that some will be lost irretrievably. Conservation of the historic environment in England originated with the establishment of the Society for the Preservation of Ancient Buildings by Morris and his colleagues in 1877. That great man of Essex and his friends focused on preservation of what actually exists, and opposing contemporary changes to historic buildings. This attitude still leads to a lingering conservation approach in which change is generally perceived as negative. However, on the coast such considerations will be often be overwhelmed by natural processes that mean physical conservation frequently *cannot* be achieved. Developing a record of what we will lose *might* be possible, supposing that funds were available to do so.

Understanding coasts fully, though, involves accepting and acknowledging the dynamism that shaped them, and appreciating that many coastal historic structures and sites were always

on a static trajectory towards destruction as the landscape around them changed through time. There is even a word for it – perhaps inevitably in German – 'Vergangenheitsbewältigung', which means coming to terms with the past and the consequences for the present, whether you like them or not. This general attitude has been termed 'anticipatory history' by DeSilvey *et al* (2011), who quote William Cronon: 'Our ability to project ourselves into the future ... is merely the forward expression of the experience of change that we have learned from reflecting on the past'. Nevertheless, accepting loss, and salvaging what information can be salvaged, does not always fit easily with traditional conservation values.

Climate change

There is abundant evidence for rises and falls of eustatic (global) sea level in accordance with climate (*see* Chapter 3). During the latest, Devensian, glacial stage sea levels were more than 100m below those of today; whereas during the Ipswichian interglacial they were some 5–6m higher than those of today. In more recent times, sea level rises (around 20cm in the 20th century) have resulted overall from thermal expansion of the oceans, combined with melting glacier and polar ice. At a temperature of 5°C a rise in water temperature of 1°C results in an increase in volume of around 1 part in 10,000, which is trivial in terms of a teaspoon, but of vast significance when considering our oceans.

As climate change proceeds, further sea-level rise is to be expected, though its rate and scale are uncertain. Our current understanding is based on projections derived from models, based on the data available, which may be inadequate. The specific impacts and rate of climate change are uncertain. The one thing about which we can be confident is that change will happen. Projections of the absolute amount of sea-level rise depend on the scenario being considered. On current models, polar ice sheets in Antarctica appear close to balance, in terms of snowfall and ablation, though in Greenland losses are occurring. Already observations from satellite radar altimeters show increases in ice-mass loss in Greenland (90km^3 in 1996; 140km^3 in 2005, with still greater losses in 2012), acceleration in movement of coastal glaciers (Fig 6.1), and penetration of melt-water to the bed of ice-sheets where it acts as a lubricant, accelerating movement. The University of Washington's Pan Arctic Ice Ocean Model Assimilation System (PIOMAS) estimates that sea ice volumes fell in late August 2012 to roughly 3,500 cubic kilometres – 72 per cent lower than annual means over the period 1979–2012. In the longer term, melting of ice on Greenland could result in a global sea-level rise of 7m, though the time-scale over which this might operate is very uncertain. There also remains the possibility of much more rapid, singular, non-linear change: for example dis-integration of the West Antarctic Ice Sheet. This is grounded below sea level and hence vulnerable to ocean temperature change: its loss would add 4–6m in sea level (Houghton 2009, 176–81 and 229). It is possible to consider

managing the coastal historic environment given the less extreme scenarios but in the unlikely event that the *worst* happened society would have more pressing concerns to consider, related to simple survival.

Sea-level change is the underlying factor in determining the forms of coasts, in combination with tectonics, isostatic change, geology, topography, storminess and wave climate. Moreover, since the oceans take time to warm, residual effects can be anticipated even were emissions of greenhouse gases stabilised. However, some of the most dramatic changes in coastlines occur during short-lived weather events, especially storms and storm surges, which are also (though less certainly) projected to increase. Precipitation will certainly rise overall, as the water-holding capacity of the atmosphere rises in accord with temperature rise (by about 6.5% per degree C). The atmosphere will become more energetic, owing to increased releases of latent heat as more water vapour condenses. Future planning for the coastal historic environment therefore has to consider the likelihood of more flooding in low-lying areas, and increased rates of erosion, as beaches are over-topped, permitting direct impact of waves on the toes of dunes and cliffs.

Figure 6.1
Jokulsarion, Iceland. Small icebergs on the beach. Bergs of this type transported erratics to the channel coasts of England during glacial stages.

More specific changes for the coastal and marine environment of the UK are discussed in Jenkins *et al* (2009a; 2009b), in Lowe *et al* (2009) for the UK Climate Projections 2009 (UKCP09) and in Defra (2012), for the UK Climate Change Risk Assessment. Tide gauges and satellite data indicate a considerable rise in global sea levels since the mid-19th century, with an increased rise in the last 20–30 years. Mean sea levels are already rising by about 3mm per year, and the UKCP09 projections indicate an estimated mean sea-level rise of between 13 to 76cm by 2095. However, because of isostatic and tectonic changes in land levels this converts to a relative rise of approximately 21 to 68cm for London and 7 to 54cm for Edinburgh (5th–95th percentile for the medium emissions scenario). These estimates, however, are thought to be conservative. The UKCP09 includes an additional H+++ scenario which is 'beyond the likely range but within physical plausibility' of 93cm to 1.9m by 2100. This is based largely on the maximum rise level given by IPCC AR4, and on proxy indications of sea-level rise in the last interglacial.

Rising relative sea level affects all coasts but its impacts are affected by other natural variables, and on whether coasts are defended artificially or not (Fig 6.2). Some 30 per cent of England's coast is eroding, but it is also a heavily defended coast (46 per cent). Beaches will be reduced, and salt-marsh, vegetated shingle and coastal sand dunes will roll inland, unless constrained by sea defences. Cliff erosion will continue and soft cliffs of till, clays and shale will erode more rapidly.

Of the other climate variables considered in UKCP09, the most significant for the coast is annual rainfall *distribution*. Central estimates (50 per cent probability) of overall annual precipitation imply little change. However, projections of changing distribution of precipitation through the year are potentially significant. By the 2080s the projections suggest increases in the winter of up to 33 per cent (50% probability: +9 to +70% at 10% and 90% probabilities) in the west of England, with slight decreases in the uplands of Scotland. Conversely, precipitation in summer might decrease by around 40 per cent in parts of southern England. Increased rainfall in winter months is likely to lead to waterlogging of soils, increased surface run-off, sudden increases in river flows, and over-bank flooding of low-lying coastal land. In urban areas the designed capacity of drains and sewers may be exceeded,

Figure 6.2
Happisburgh, Norfolk.
This was not a storm, just a
rough day with a north-east
wind, as waves beat over
failing sea defences.

Figure 6.3
Eastney, Portsmouth,
Hampshire. This Victorian
beam engine was essential
to permit large-scale
residential development
on a low-lying island.
It pumped sewage to the sea.

resulting in surface- and foul-water flooding (Fig 6.3). This has already been especially severe at some low-lying coastal towns, where gravity flow to the sea, unaided by pumping, is inadequate. In many places the main factor causing coastal land-slips is the rate of percolation of groundwater through the cliff sediments, destabilising the entire cliff-face, especially where cliffs are composed of unconsolidated material such as Jurassic clays and shales (for example in Dorset and the Isle of Wight), or Pleistocene tills and glacial outwash sands (for example in Holderness and East Anglia). There were sudden cliff collapses in late 2012 in Whitby and Burton Bradstock, the former causing losses of houses and the latter a fatality. Collapse will be accelerated by changes in seasonal rainfall distribution, especially wet winters (McInnes 2008, 33–6), with potential impacts on cliff-top historic structures and archaeological deposits.

Lowe *et al* (2009) consider some other sub-sea surface variables in shelf seas around the UK: temperature, and especially salinity and circulation. For a medium-emissions scenario the shallow seas around the UK are projected to be in the range 1.5–4°C warmer by 2100. Over the same time period salinity is projected to decrease by 0.2 practical salinity units (PSU),

while there will also be changes in seasonal salinity stratification periods. Changes in these variables would have few direct physical impacts on archaeological sites, but they would affect the distribution of marine organisms. *Lyrodus pedicellatus*, the Siamese shipworm, is apparently native to the Indo-Pacific oceans, but latterly has been reported from southern England, where it has not been seen before. It has been recorded from a timber wreck in the Swash Channel at the mouth of Poole Harbour, and in those timbers of the 16th-century wreck of the *Mary Rose* that are still submerged beneath the Solent (Palma 2009). Unlike the northern species of shipworm, *Teredo navalis*, it is active all year round and so will accelerate the destruction of any submerged or intertidal archaeological wooden structure exposed above the sea floor.

Legislation and Flood and Coastal Erosion Risk Management (FCERM)

The Climate Change Act 2008 requires the Secretary of State to ensure that the net UK Carbon Account for all six Kyoto gases for the year 2050 is at least 80 per cent lower than the 1990 baseline, toward avoiding dangerous climate change. The aim is to make the UK a low-carbon economy and gives ministers powers to introduce measures that achieve a range of greenhouse gas reduction targets. There is an independent Committee on Climate Change advising government. Following on from this Defra has produced a Climate Change Risk Assessment (CCRA) which assesses 100 risks (prioritised from an initial 700). It focuses on risk in terms of key sectors: agriculture and forestry; business; health and well-being; buildings and infrastructure; and the natural environment. Space does not permit detailed assessment of all potential impacts of risks on the historic environment but in practice there are very few risks identified in the CCRA that have no potential historic environment impacts.

This is being followed by a National Adaptation Programme (NAP) (www.defra. gov.uk/environment/climate/government/), which will focus attention on assessing the risks of climate change. The first NAP was published in 2013 and focused on helping organisations to become more resilient or 'Climate Ready' to climate change impacts. It is now being extended to include the historic environment.

The Government's policy on managing coastal change in the context of a rapidly changing climate, and assisting coastal communities adapt to that change, is underpinned by 'Making space for water: developing a new Government strategy for flood and coastal erosion risk management in England' (Defra 2004), 'A strategy for promoting an integrated approach to the management of coastal areas in England' (Defra 2008) and Sir Michael Pitt's review 'Lessons learnt from the 2007 floods' (Pitt 2008), and 'Adapting to Coastal Change: Developing a Policy Framework' (Defra 2010). The Flood and Water Management Act 2010 required the Environment Agency to 'develop, maintain, apply and monitor a strategy for flood and coastal erosion risk management in England'. The National Flood and Coastal Erosion Risk Management entitled 'Understanding the risks, empowering communities, building resilience' was released in 2011 (www. environment-agency.gov.uk/research/**policy**/ 130073.aspx). Its principles include community focus and partnership working, a catchment and coastal 'cell' based approach, sustainability, a proportionate, risk-based approach, multiple benefits, and encouragement of beneficiaries to invest in risk management.

The emphasis now is on managing risk rather than attempting to provide absolute defence, which probably never was achievable anyway. This approach is very different from that which followed the catastrophic flooding of 1953, caused by a North Sea storm surge. The official estimate of deaths in the Netherlands then was 1,835, while 307 people died in the east of England – 112 in Essex alone – and around 24,500 houses were destroyed or damaged there. Thousands of livestock were drowned, and extensive areas of farmland were contaminated with saline water (Grieve 1959). The events of 1953 defined the flood management agenda for decades on into the 20th century (Murphy 2009, 181–3). It is worthwhile to consider the prevailing mind-set of the country then, which had learnt only eight years before in World War II that endurance, and not counting the cost, eventually brought victory. Military defence and flood protection are, of course, two quite different things, but when a nation has the *perception* of being assaulted they may seem comparable to people at the time. The very terms still used, even today, in flood risk management, such as 'Hold the line', have obvious military resonances. Large sums were expended on new coastal defences, sometimes without proper appraisal of the benefit:cost ratios for the schemes, and with little or no consideration of ecological effects. It took a strange pair of bedfellows – environmentalism, and the new political and economic attitudes of the 1980s – to change attitudes towards more minimal intervention, where possible.

The main aims of coastal change policy now are to manage the risks associated with flooding and erosion in order to reduce the threat to people and their property and to deliver the greatest environmental, social and economic benefit, consistent with sustainable development principles. Risk is conventionally defined as a function of the *probability* of an event happening, and the *consequences* if it did. Sites and monuments of all types, and of all periods, are now in coastal locations that are, to a greater or lesser degree, threatened by damage or loss as a result of on-going coastal change. However, both probabilities and consequences are very variable. There is no certainty: it is probabilistic.

One major output of the *National Heritage Protection Plan (NHPP)* is information that will aid enhanced designation, and hence conservation, of historic assets. Only a very small proportion of recognised and recorded historic assets – less than 5 per cent – have any form of statutory protection. This is especially true on coasts: there *are* designated sites, including those investigated in English Heritage's CERA (Hunt 2011), but a high proportion of sites that can potentially be designated have no protection. Many more are in categories that cannot be designated under existing legislation, such as artefact scatters.

The National Planning Policy Framework (NPPF: www.communities.gov.uk/planning andbuilding/planningsystem/planningpolicy/ planningpolicyframework/) incorporates the principles of earlier planning guidance and statements, requiring desk-based assessment, field evaluation and, if necessary, 'Local planning authorities should … require developers to record and advance understanding of the significance of any heritage assets to be lost (wholly or in part) in a manner proportionate to their importance and the impact, and to make this evidence (and any archive generated) publicly accessible' (paragraph 141). This is obviously applicable to new commercial coastal developments such as those related to ports or leisure. It also applies to coastal management schemes, such as Managed Realignment, approved and partly or wholly funded by the Environment Agency. As specified in the Environment Act 1995: 'It shall be the duty of each of the Ministers and of the Agency, in formulating or considering … any proposal relating to any functions of the Agency …(i) to have regard to the desirability of protecting and conserving buildings, sites and objects of archaeological, architectural, engineering or historic interest; (ii) to take into account any effect which the proposals would have on the beauty or amenity of any rural or urban area or on any such flora, fauna, features, buildings, sites or objects …' (chapter 1, section 7).

This is being implemented by the Environment Agency (2007), but in a defined way: 'In cases where the DBA and coastal modelling indicate an increase in the threat to the historic environment as a direct result of the Environment Agency's engineering works (managed re-alignment), the Agency will seek to carry out site investigation of known heritage assets, and assess the potential for the survival of currently unknown remains. The Environment Agency (EA) will not investigate historic environment assets, known or predicted, within the re-aligned area i.e. the area proposed to be reclaimed by the sea, unless they are directly affected by our engineering works or construction activities'. In short, the EA accepts responsibility as a developer for direct constructional impacts on historic assets, but not for indirect effects, such as submergence. In practice, however, EA funding has been generous at several recent Managed Realignment Schemes, and has ensured that understanding of the development of the changing landscape is achieved (*see below*). The Agency deserves congratulation. Natural processes of coastal change are plainly not considered. Where historic assets are affected by natural processes of erosion or flooding planning regulations do not apply, and hence there is no obvious source of funding to ensure recording before loss.

Climate change planning involves mitigation of emissions and adaptation to change. EH has strategies in place to mitigate carbon emissions from its operations and has provided guidance on flooding and historic buildings and on their adaptation for sustainable energy production (English Heritage 2008b; 2010b; 2011b). As part of the EH National Heritage Protection Plan (NHPP), projects include work on the thermal performance of traditional building elements; whole-house thermal performance and the impacts of interventions; research into the technical risks of insulation; and work on improving energy models for traditional buildings (Fig 6.4). Following recent restructuring, a Climate Change Officer has been appointed in the Historic Environment Intelligence Team, and the development of an EH Climate Change Network to provide responses to Government Climate Change initiatives, to revise the current EH Guidance *Climate Change and the Historic Environment* (English Heritage 2008c), and to contribute towards EH's report to Defra on adaptation. Other EH Guidance will be produced, including revision of *Coastal Defence and the Historic Environment* (English Heritage 2003b). Where nationally important sites are threatened with destruction by coastal change and require excavation, EH is in principle the body responsible for providing funding: in practice this may be limited at a time of financial austerity.

Figure 6.4
Fort Cumberland, Eastney,
Portsmouth. Micro-
generation at a property
managed by English
Heritage.

Shoreline Management Plans (SMPs)

SMPs provide a large-scale assessment of the risks associated with coastal processes and present a long-term policy framework to reduce these risks to people and the developed, historic and natural environment in a sustainable manner. An SMP is a high-level document that forms an important element of the strategy for flood and coastal erosion risk management. Coastal Groups, made up primarily of coastal district authorities and other bodies with coastal defence responsibilities, provide a forum for discussion and co-operation and play an important part in the development of SMPs for their area (www.defra.gov.uk/Environ/Fcd/guidance/smp.htm). Environmental appraisal for SMPs is required by the 'Strategic Environmental Assessment (SEA) Directive' (2001/42/EC) and, to this end, SMPs include Thematic Reviews identifying the key features of the natural, human, historical and landscape environments of the coast with a commentary on their characteristics, status, designations, their importance and the benefits they provide to wider society. The Environment Agency has a national strategic overview for SMP production. SMPs are intended to be reviewed approximately every ten years, and the most recent stage has just been completed. Guidance for SMP review is given in Defra (2006). English Heritage has published

Shoreline Management Plan Review and the Historic Environment: English Heritage Guidance (2006) to supplement the Defra guidance with respect to heritage assets.

The coasts of England and Wales have been divided into 11 Littoral Cells – extensive lengths of coast (for example Cell 3 – Wash to Thames), within which sediment moves, but between which sediment transport is considered to be minimal. Each cell is the responsibility of Coastal Groups, led by the Environment Agency. The cells are further sub-divided into more manageable units known as sub-cells. An SMP sets the future policy for cells and sub-cells, still further sub-divided into shorter lengths of coast, termed Policy Units. An SMP should 'provide the basis for policies for a length of coast and set the framework for managing risks along the coastline in the future' and 'identify the best approach or approaches … over the next 100 years'. The SMP does not consider how these policies should be implemented: that is dealt with in subsequent stages of strategies and individual schemes. 'Hold the existing defence line', as the term suggests, will involve maintaining, and often improving, existing defences. 'Advance the existing defence line' will relate to situations where new land reclamation is the best option, although it seems unlikely that it will be used widely. 'Managed Realignment' calls for the identification of a new sustainable coastal defence line and construction of new defences landward of the existing defences. Finally, 'No active intervention' means just that – allowing dynamic coastal processes to proceed in an unconstrained way, with no investment in defences or, alternatively, to allow existing defences to deteriorate or be over-topped without maintenance or repair.

The latter two options will almost always have impacts on the historic environment. 'Managed Realignment' involves breaching sea defences, (some of them of considerable antiquity: *see* Chapter 4), while construction works for the new sea wall and/or new drainage systems for the realigned area, may cut through, and damage, archaeological sites. The effects of re-wetting buried sites with saline water are hard to assess without experimental evidence but, very likely, will lead to disintegration of poorly fired ceramics, enhanced corrosion of metal artefacts, and changes in microbial activity leading to decomposition of organic materials that were in a semi-stable state. 'No active intervention' just permits continued erosion, with consequent physical loss of sites, for example by the development of gullies cutting down into Holocene sediments, cliff erosion and dune retreat. The policy options cover time spans of 20, 50 and 100 years, and are subject to review approximately every decade. Consequently the preferred option may change through time.

Placing historic assets within the process of Shoreline Management depends, crudely, on scoring their significance. Designated assets plainly have a high priority but many other assets do not fall within the criteria permitting designation. They are discussed below. Criteria to be considered in assessing significance and determining management would be Threat, Condition, Significance, Potential for providing new information, Rarity, and possibly Potential for Designation: *see*, for example, in NE2 and also Department for Culture Media and Sport (2010).

Assessing risk

English Heritage has produced a Coastal Estate Risk Assessment (CERA), looking very specifically at the risks in the 21st century for the coastal historic properties which it manages. The study examined 54 EH coastal properties that fulfil the criteria of the study (Hunt 2011). These properties, with their associated infrastructure have been assessed against Environment Agency flood and coastal erosion mapping data. The analysis and data collation was done in a GIS, to facilitate comparison of the available datasets. All properties had some level of potential threat, but of the properties assessed two were considered to be at high risk of flooding (Landguard Fort (Fig 6.5) and Berney Arms Windmill) and four at high risk from coastal erosion (Daw's Castle, Garrison Walls and Innisidgen Burial Chambers). Elsewhere some sites are

Figure 6.5
Landguard Fort, Suffolk.
The site is designated but its setting has been extensively modified by expansion of the Port of Felixstowe.

sustainable only because of existing hard coastal defences (for example Reculver – Fig 6.6) or beach management (for example Hurst Castle). Were these measures not continued the sites would be at high risk. The report includes recommendations for future management. Of course, CERA has been focused on a limited number of high-profile EH Historic Properties. As part of the NHPP a study to assess flood and erosion risk for inland flooding has been initiated. More generally there are wider risks, which need to be considered in relation to specific types of coastline.

Dune coastlines

Fulford *et al* (1997, fig 31) give a national map of dune coasts, which in general occur in the North-East, the east of England, the South-West and North-West, with some sites in the south. Dunes are inherently dynamic and mobile, responding to changes in relative sea level and beach levels, though they may temporarily be stabilised by growth of vegetation, initially by growth of marram grass (*Ammophila arenaria*) and then by more diverse vegetation including other grasses (for example *Festuca rubra*), sedges (for example *Carex arenaria*), and herbs to become 'grey' dunes (jncc.defra.gov.**uk**/protectedsites/

sacselection/habitat.asp??featureintcode`=h2130). In the past, and often today, they have determined the positions of habitable coastal areas by isolating areas of land at or below mean sea level from the sea itself. The chronology of dune development is often not well understood, but it seems probable that mobility was enhanced during stormy climatic phases, such as the 'Little Ice Age', around 1550–1850. Some dunes are calcareous, including shell fragments, whereas others are acidic, which strongly influences the types of palaeoecological indicators (mollusc shells, bone, pollen) that survive within them.

In terms of archaeological site management, dune mobility and erosion of seaward faces may be critical for sites that lie beneath or within dunes. Inland migration of dunes may result in sites formerly landwards of them being exposed to direct marine erosion. The Bronze Age site of Seahenge at Holme-next-the Sea was originally constructed landwards of a dune system; now, due to long-term dune retreat over the last 4000 years, it lies on the beach (Murphy and Green 2003, fig 47). At Low Hauxley on the north end of Druridge Bay, Northumberland, archaeological material (including a Mesolithic flint scatter and a Bronze Age cist cemetery) lies beneath a dune system (Fig 6.7). Here there is an unusual form

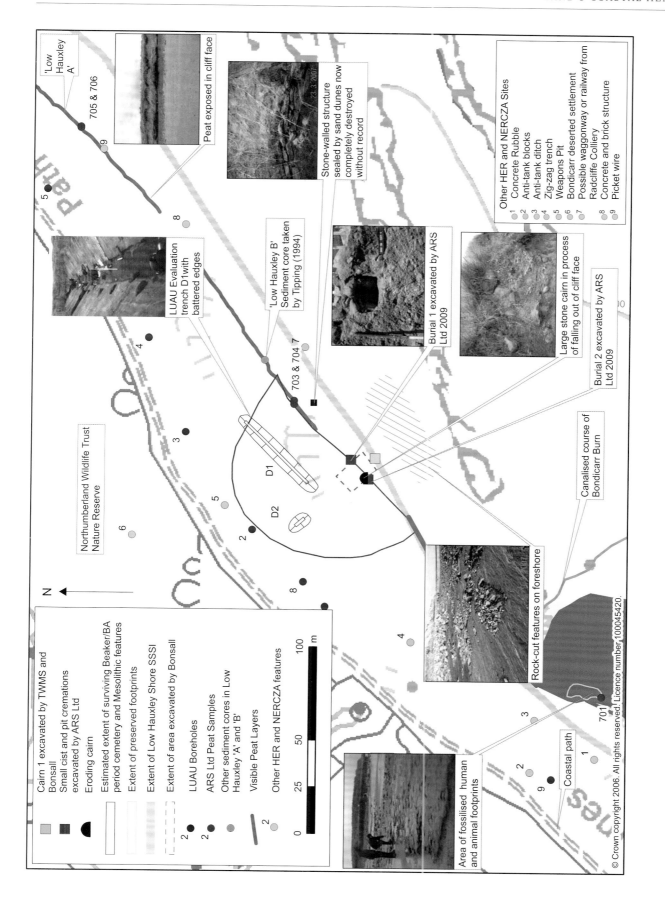

'Low Hauxley A'

705 & 706

Peat exposed in cliff face

Stone-walled structure sealed by sand dunes now completely destroyed without record

Other HER and NERCZA Sites
1 Concrete Rubble
2 Anti-tank blocks
3 Anti-tank ditch
4 Zig-zag trench
5 Weapons Pit
6 Bondicarr deserted settlement
7 Possible waggonway or railway from Radcliffe Colliery
8 Concrete and brick structure
9 Picket wire

LUAU Evaluation trench D1 with battered edges

'Low Hauxley B'
Sediment core taken by Tipping (1994)

703 & 704

Burial 1 excavated by ARS Ltd 2009

Northumberland Wildlife Trust Nature Reserve

Large stone cairn in process of falling out of cliff face

Burial 2 excavated by ARS Ltd 2009

D1

D2

Canalised course of Bondicarr Burn

Rock-cut features on foreshore

N

Cairn 1 excavated by TWMS and Bonsall
Small cist and pit cremations excavated by ARS Ltd
Eroding cairn
Estimated extent of surviving Beaker/BA period cemetery and Mesolithic features
Extent of preserved footprints
Extent of Low Hauxley Shore SSSI
Extent of area excavated by Bonsall
LUAU Boreholes
ARS Ltd Peat Samples
Other sediment cores in Low Hauxley 'A' and 'B'
Visible Peat Layers
Other HER and NERCZA features

0 25 50 100
m

Area of fossilised human and animal footprints

Coastal path

701

of 'coastal squeeze' as there has been open-cast coal mining landwards while the seawards face of the dune system is eroding (NE5, 162–5). The dune system thus has no space in which to re-form. Well-preserved World War II military features are likewise vulnerable to dune recession and erosion at this location (NE5, 145–8).

Salt-marsh, marshes and estuaries

Salt-marsh supports highly significant natural communities of plants and animals and also provides 'ecosystem services', primarily in terms of reducing wave energy so that artificial tidal defences behind the marsh have a degree of protection (Environment Agency 2011). Salt-marsh sediments include, or overlie, archaeological deposits and so understanding the loss or growth of salt-marsh is plainly significant. Between 2006 and 2009 data obtained primarily by aerial photography provided a new national map of the extent of marshes, following on from a previous survey of 1989. Despite plain evidence that salt-marshes are being lost to erosion at some localities, especially in the south and south-east of England (Fig 6.8), accretion has been observed elsewhere. The main destructive factors have been 'coastal squeeze', where the marsh is confined between the sea and hard coastal defences and thus exposed to the effects of isostatic change, sea-level rise and increased storminess. Under natural conditions it would re-form further inland, but artificial defences prevent this. The current estimates suggest that salt-marsh loss may be 100ha per year or less, which is slower than anticipated. In terms of achieving compliance with EU Directives (Habitats, Birds, and Water Framework) this is welcome news, overall, in terms of nature conservation. In archaeological terms some sites will be protected by accretion, but in other places will be lost. It is the specific location of loss that is of key concern.

Besides physical loss of marshes, Tomalin et al 2012, 438–84), referring specifically to the Isle of Wight, discuss the numerous factors that affect intertidal and sub-tidal sites, including the activities of bacteria and marine boring organisms (which directly influence wood preservation), physical erosion, bait digging, trawling, shipwash, and dredging for shellfish or for navigational purposes or minerals.

Many salt-marshes have been converted to grazing marsh by construction of sea defences

Figure 6.8
Sutton, Suffolk. Salt-marsh erosion has led to direct wave impact on the sea wall, which is now failing.

Figure 6.9
Blakeney Marshes, Norfolk.
Excavation of a medieval
structure (the so-called
Blakeney Chapel) prior to
realignment of the mouth
of the Glaven Estuary.

(*see* Chapter 3 and Figs 6.9–6.10). Coastal marshes are widespread in Suffolk and were assessed as part of the survey (S3, 24–35). In general terms late reclamation is represented by a very regular pattern of dykes, whereas pre-1600 land-claim is marked by serpentine dykes, frequently following former salt-marsh channels, though more complex multi-phase reclamations can also be seen. Widespread conversion to arable or to grazing 'improved' by use of artificial fertiliser accelerated, and by the early 21st century only about 3 per cent of the

Figure 6.10
Freckleton, Lancashire.
Grazing marshes include a
range of fossil natural and
anthropogenic features.

former 10,000 hectares consists of unimproved grazing (Williamson 2005, 27–49). Elsewhere, changing agricultural practice in the 20th century resulted in a 25 per cent loss of the Lincolnshire Coastal Grazing Marsh to arable production between 1990 and 2000, and the loss of around two-thirds of grazing marsh in the Thames Estuary between the 1930s and 1980s. Natural England has initiated the 'Coastal Grazing Marsh Project' in Lincolnshire, and the Essex and North Kent marshes have been designated as Environmentally Sensitive Areas, with the aim of arresting and, ideally, reversing this trend (Tann 2004; Wetland Vision Project 2008). By the end of the 1990s there were around 6500 hectares of surviving coastal grazing marsh in Essex, which represents some 5.5 per cent of the national resource (Essex Biodiversity Project 1999). Countrywide, archaeological earthworks, including field systems, settlement sites and saltern mounds, have been destroyed slowly by repeated ploughing, or abruptly by intentional bulldozing. An unknown number of archaeological sites has been lost. The lowered water-tables have also led to the drying out of formerly waterlogged sites, so wooden structures and other organic materials would have undergone biogenic degradation.

Sites at low elevations around estuaries are, in addition, vulnerable to freshwater flooding owing to prolonged and intense periods of rainfall combined with problems of water delivery (English Heritage 2010a). The Roman palace at Fishbourne, near Chichester, was constructed in about AD 75 on a vast raised platform of clay and gravel at elevations of around 4.57m to 6.1m OD. This raised the new building above contemporary tidal limits, though still with high saline groundwater. The palace includes some of the finest and earliest mosaic floors in Britain (Cunliffe 1971; 2010). During the exceptionally rainy summer of 2012, the wettest since 1912, parts of the site were flooded (Fig 6.11). The immediate cause seems to have been a leaking water main, but the site remains vulnerable to rising groundwater over the longer term. Besides the potential effects on the physical integrity of the mosaics by degradation of mortar, once the water receded, algae and other micro-organisms developed on them. It has proved necessary to use ultra-violet light to attempt sterilisation of the surfaces. Historically, catastrophic sea-flooding is marked at St Nicholas, New Romney, Kent, where staining on the nave pillars and a raised ground level outside the church mark deposition of flood sediments in 1287 (Fig 6.12).

Seawards of defences in estuaries, historic assets are exposed and vulnerable to marine erosion. For example, at Amble, Northumberland, there is a hulk assemblage comprising timber vessels thought to be wherries used as

Figure 6.11
Flooding at Fishbourne
Roman Palace, Sussex,
in 2012.
(Image by Robert Symmons
with permission from
Sussex Past)

Figure 6.12
St Nicholas, New Romney,
Kent. The Norman doorway
is well below modern street
level owing to deposition of
flood deposits in the late
13th century.

coal lighters and fishing vessels. Continued degradation by physical erosion and marine boring organisms is inevitable (NE5, 187–8).

Cliff coastlines

Cliffs continually erode, though the rate of loss is determined primarily by beach levels, the geology of the cliffs, their exposure to the prevailing wave climate, and freshwater flow through them. Readily eroding cliffs are composed of glacial till, sands and gravels (mainly in the east of England) and shales and clays of Jurassic date (mainly in Dorset), but a degree of erosion is prevalent even along cliffs of hard igneous rock. Moreover, sites located on unstable substrates may be prone to long-term movement and occasional major landslides, as at Ventnor on the Isle of Wight (McInnes 2008). However, erosion and coastal landslides do not necessarily always represent just loss: Palaeolithic sites stratified within cliff sediments may be exposed. Gisleham Cliff, Pakefield, has produced lithics of pre-Anglian date (Parfitt et al 2005).

However, cliff-top sites are often at risk, and where the cliff-face is unstable a substantial zone of a site may in effect be archaeologically sterilised since it is too hazardous to undertake excavation very close to the cliff edge. The North Yorkshire coast between Whitby and Flamborough is characterised by resistant rock headlands, such as at Robin Hood's Bay, Whitby and Flamborough Head, with intervening stretches of cliffs capped by till (glacial clay) and shore platforms (YL2, 17–20). Mean rates of erosion are around 0.25m per annum for the till cliffs, but they are prone to sudden catastrophic rotational failure, involving slippage of large semi-circular areas of land, especially when groundwater is suddenly replenished by prolonged rainfall after drought. A recent event involved the loss of the Holbeck Hall Hotel at Scarborough in 1993 but, historically, large areas of land and historic assets have been lost. Filey Brigg, for example, consists of a narrow ridge of till over rock, and here a Roman signal station was excavated in 1993–4, prior to its loss by erosion (Ottaway 2001).

Erosion is also a problem on rocky headlands, notably at Whitby, where excavations were undertaken by English Heritage in 1993, and more recently in 2007, to record features associated with the abbey and associated Anglian settlement before their loss. The erosion of the headland is being monitored by LiDAR survey, combined with historic map regressions. The overall rate of erosion is 0.22m per annum, although significantly greater rates are being observed in some areas. This gives some means of predicting areas of known or suspected archaeology requiring excavation prior to their loss (Miller *et al* 2008). Cliff collapses within the town of Whitby in 2012, resulting in the loss of buildings, were a consequence of instability caused by freshwater flow through the rock, again following an unusually wet summer.

Further south, the Holderness coast consists mainly of low cliffs of till, which is easily eroded. Localised beach depletions known as ords enhance cliff erosion rates, and the ords migrate southwards at about 0.5m per annum, from Barmston to Spurn Head, which has resulted in a mean annual loss of around 150m since the publication of the 1st edition Ordnance Survey maps in the 1850s, but earlier erosion resulted in the loss of some thirty towns and villages since the Middle Ages (Sheppard 1912). On the south coast, the eroding Roman villa and pre-existing Iron Age occupation – conceivably representing an early trading site or *oppidum* – on a chalk cliff at Folkestone, Kent, is eroding and is currently under excavation (Selkirk 2012).

The archaeological solution to cliff erosion is, plainly, excavation and recording before loss. However, funding to achieve this may be problematic. Volunteer participation may often be one solution (*see below*).

Figure 6.13
St Cuthbert's Island,
Lindisfarne,
Northumberland. Medieval
features here are eroding.

Islands

Islands vary in character from isolated rock outcrops through to former islands in mud-flats, now represented by undulations of underlying soft geology, often partly sealed by alluvium and visible as hummocks within reclaimed land. The Premonstratensian Abbey, originally of 1182, at Leiston, Suffolk, was relocated inland from its original site, which was on a low island in the marshes but now lies in grazing marsh. Some ruins of the original abbey survive, with clear cropmarks defining its lay-out. Proposals for Managed Realignment here would reinstate the site's original island location (S3, 48). At Burrow Hill, Suffolk, an 8th-century settlement and cemetery has been recorded on a former island within salt-marsh on the Butley River. The all-male character of excavated burials could indicate a monastic community. Recent aerial photographic survey has defined cropmark enclosures in the vicinity, and an earthwork causeway known as The Thrift approaches the site (S4, 63–4). Such former islands within grazing marsh would plainly have been isolated and readily defensible foci for settlement and other activities. The main threats to them come from reclamation and/or Managed Realignment, resulting in flooding. On hard rock coastlines, by contrast, islands are more vulnerable to erosion and cliff collapse. An example is a Mesolithic site at Ness End, Holy Island, Northumberland, where there is a lithic scatter, vulnerable to scour and quarry collapse. At St Cuthbert's Island a medieval chapel and related structures have been recorded, all actively eroding (NE5, 224–5) (Fig 6.13).

Ports

Historic ports in general are undergoing change, though largely driven by economic development. At Dover, for example, freight vehicle traffic is projected to rise from 2.3 million units in 2003, to 3.12 million units in 2014, to 3.52–3.92 in 2024 and 3.72–4.52 in 2034 (Dover Harbour Board 2007). To cope with this expected demand, Dover, like many other ports, is intending to expand: a new terminal at the Western Docks is proposed (Fig 6.14). Proposals for port expansion are at various stages of planning, development and completion elsewhere, for example at Harwich (Bathside Bay), Felixstowe (South Extension),

Figure 6.14
Western Docks, Dover, Kent.
Development of a new
terminal at the port is
resulting in remodelling
of the harbour, including
reduction of a pier and
infilling of a harbour basin.

Sheerness, Avonmouth and Liverpool. A completely new port for London, the London Gateway, on the site of the old Shellhaven Oil Refinery is under development.

Where these developments are at preexisting ports, the impacts are partly on existing historic port buildings (including warehouses, Customs Houses, workshops etc), quayside structures and potentially other elements of port infrastructure, including lighthouses and lifeboat stations. There may, however, be 'heritage gains' in terms of restoration, and new uses for historic buildings that, at present, have no function and are dilapidated (Fig 6.15).

Figure 6.15
Sefton Street, Liverpool.
An 18th-century house
and warehouse.

Figure 6.16
St Mary's, Isles of Scilly.
Erosion of superficial
deposits over granite
threatens defensive
structures of the
Garrison Walls.

Where areas of foreshore or shallow subtidal water are being reclaimed to construct new berths, archaeological sites may be damaged or destroyed by engineering works and compaction. New capital dredging, to create larger channels to accommodate the massive vessels now in use could potentially have adverse impacts on historic wrecks and other submerged sites. However, Port Authorities increasingly accept the necessity of diverting new navigation channels around the most significant wrecks, and recording others prior to dredging. But port expansion may also have knock-on effects. Frequently ports are in estuaries and are surrounded by areas of designated wild-life habitat, including Special Protection Areas (SPAs) and Special Areas of Conservation (SACs), which require compensation for any habitat lost as a consequence of development. This necessitates creation of new areas of equivalent habitat by Managed Realignment, with effects on any archaeological sites within the realigned areas. In fact these need not necessarily be close to the development area at all.

Military defences

Coastal military defences were, in general, sited directly on their contemporary coastline and consequently are especially vulnerable to erosion. This is especially true for the 'coastal

crust' defences of the two World Wars (*see* Chapter 4). Many substantial concrete structures have been undermined by erosion, or have fallen from cliffs and now lie displaced or inverted on beaches. The Garrison Walls on the Isles of Scilly comprises a nationally important example of a multi-phase coastal fortification, constructed and modified from the 16th–20th centuries. Parts are threatened by coastal erosion (Fig 6.16). Already, collapses have necessitated repair and reconstruction, as at the Lower Benham Battery (Bowden and Brodie 2011, fig 85). However, continued interventions of this type cannot be continued indefinitely and losses of parts of the defensive circuit are inevitable eventually. Archaeological survey, excavation and photogrammetric recording have been undertaken to provide an accurate record of the monument for posterity and to provide a baseline record against which further deterioration can be monitored.

Historic Environment Designation

The purpose of designation is to help protect historic assets (DCMS 2010). Designated assets include: Scheduled Monuments (SMs) designated under the Ancient Monuments and Archaeological Areas Act 1979; historic shipwrecks designated under the Protection of Wrecks Act 1973; and Listed Buildings and Conservation Areas designated under the terms of the Town and Country Planning Act 1990. Listed buildings are graded I, II*, or II. Decisions on works affecting scheduled monuments and historic wreck sites generally require a specific permission from the Secretary of State for Culture, Media and Sport (DCMS) alongside any other consents such as planning permission. Works to listed buildings may require Listed Building Consent, and the demolition of buildings within a conservation area may require Conservation Area Consent. Other historic sites, including World Heritage Sites, historic parks and gardens and historic battlefield sites are included within non-statutory registers, which underline the need to consider their special importance within the planning process, when development is proposed. Designation of Maritime and Naval historic assets is considered in English Heritage (2012b) placing emphasis on scheduling within territorial waters, threatened sites, environmental evidence, harbours and docks.

The Rapid Coastal Zone Assessment survey programme was not initiated in the 1990s with designation specifically in mind. The earlier reports make little reference to it, though the potential for designation is addressed in the more recent ones. Designating a coastal historic asset that is actually or potentially under threat from natural erosion or flooding does not, in itself, enhance the conservation of the asset, unless funding for protection against coastal change or recording can be obtained. However, designation will serve another function in terms of highlighting the special interest of the asset. The first generation of Shoreline Management Plans was very inadequate from the historic environment viewpoint, usually referring *only* to designated assets. The recently completed second generation (SMP2) is more comprehensive: many of the plans refer to all known historic assets. Nevertheless, designations remain very significant in terms of influencing decisions on coastal management. For example, the designated Roman fort and church at Reculver, Kent, are currently protected by cliff armouring of rock rubble, and Hurst Castle, Hampshire, is partly protected by nourishment of the shingle spit on which it is located (Fig 6.17). These defensive measures are based on

Figure 6.17
Hurst Castle, Hampshire.
Sea defences with a partially ruined structure on beach.
(Photo: Abby Hunt)

the sites' historic and landscape significance. The sites are safe as long as funding is available for such defence, but it is doubtful whether similar measures for protection would have been put in place for non-designated sites.

Another approach to managing the coastal historic environment is provided by Historic Seascape Characterisation (HSC). Unlike designation, it does not impose any sense of *value*. Instead this programme aims to define areas of seascape character. Unlike Historic Landscape Characterisation it is three-dimensional, including the sea bed, the water body, and the sea surface. Furthermore, unlike land-based perceptions it is less based on visual experience, rather more on remotely or historically acquired information. It simply defines how areas of sea have been used. It is particularly well suited for wider marine planning, which is now being developed by the Marine Management Organisation. It is marine-based but the extent of HSC inland above the Mean High Water is variable: it can include reclaimed land, such as grazing marsh, or inland features such as church towers that have had a secondary use as sea-marks for mariners (www.english-heritage.org.uk/characterisation).

Managing change on a site-by-site basis

We know the starting point for a vulnerable historic coastal building or archaeological site from its present condition. We also know what is often likely to be the ultimate end-point: a scatter of unstratified artefacts and rubble on the beach. How do we move from one to the other in a controlled way? More specifically how do we manage this change so as to ensure that information of significance is not lost? There are difficulties. For example, there are some coastal buildings for which EH or other bodies provide funding for maintenance and repair and which will eventually be at risk from erosion. At what point should the decision to be made that the site is unsustainable and further investment is fruitless? Problems of this sort are still under consideration, but some approaches are already obvious.

First, where a structure is at risk from sea-flooding, measures can be put in place to enhance its resilience. Adaptive measures can include localised but permanent flood barriers, demountable barriers, temporary air-brick covers and door-boards. Resilience measures for

Figure 6.18
Dial House, Brancaster,
Norfolk.

historic buildings also include relocating electrical circuitry to above anticipated flood levels and the replacement of flooring materials vulnerable to flooding with more robust materials, easily hosed and brushed down (Fig 6.18). Essential and vulnerable facilities and equipment can be confined to upper floors. This approach has been implemented by the National Trust at Dial House, Brancaster Staithe, Norfolk, now serving as the Millennium Activity Centre and comprising Grade II Listed buildings, including the former Victory pub.

Secondly, where erosion is the threat, an asset can be relocated to a sustainable location further inland. The logistics, practicality and cost of relocation and the implications for the heritage values of the asset play an important part in decision making. Relocation is likely to be most feasible for smaller and more portable historic structures. It has, however, already been undertaken for more substantial structures threatened by coastal erosion at Beachy Head, East Sussex (a lighthouse), and Kimmeridge Bay, Dorset (a 19th-century folly and observatory) (Fig 6.19). The latter gained support from its literary associations with Thomas Hardy and P D James. Relocation will never be a cheap option, and there is dispute as to whether a relocated building is any longer historic, but rather a 21st-century building. Nevertheless, if this is the only viable option

other than abandonment it should at least be considered. The possibility of recovering project costs eventually by converting the structure for holiday rental, as at Kimmeridge Bay, should also be borne in mind.

Thirdly, assets can be recorded ahead of their loss. Archaeologists are accustomed to this process as a part of mitigation during planning: excavation in advance of commercial and infrastructure redevelopment is common-place. Buildings, however, are more prominent and may have significant local associations over and above their wider heritage significance. Abandonment (and almost certainly demolition to avoid leaving an unsafe structure) raises questions that have not yet been resolved.

Mitigation at Managed Realignment Schemes

Funding and supervision of archaeological work by the EA or Ports Authorities at Managed Realignment Schemes addresses the fundamental necessity to accept that change will happen. The existing landscape, often low-grade grazing marsh, originated as a result of human activities and the new change is just another new chapter in its history. What needs to be done is to record the landscape, and sites in it, to gain understanding of its development.

Figure 6.19
Clavell Tower, Kimmeridge Bay, Dorset, relocated from eroding cliff face.

Extensive excavation of a middle Iron Age to Roman saltern site at Stanford-le-Hope, Essex, has been noted in Chapter 4. It was investigated in advance of ground-level reduction, sea-wall construction and breaching, as ecological mitigation for development at the London Gateway Port Development (Biddulph *et al* 2012). At the Scheduled Monument of Blakeney Chapel, Norfolk, excavation funded by the EA in advance of realignment showed that the site had a long history, producing pits with Neolithic pottery, an Anglo-Saxon 6th-century gold bracteate brooch (Fig 6.20), and evidence for early medieval iron-working, with artefactual material of most intervening periods. The 'chapel' proved to be a late medieval domestic building with fireplaces, possibly a Warrener's Lodge (Lee 2005). Current unpublished work at the realignment at Steart Point, Somerset, has reconstructed the former pattern of tidal creeks (palaeochannels) underlying the modern grazing marsh. Creek-side sites, established in the Iron Age and Roman periods for salt production and farming, were followed by successive reclamation in the Middle Ages and the development of 'moated' farmsteads. Occupation in later prehistory appears to have been intermittent though in the Roman period more permanent settlements imply the existence of flood defences. At the Medmerry, Sussex, Managed Realignment Scheme excavation in advance of groundworks has revealed early Neolithic features, a middle–late Bronze Age well, burnt mound and cremation cemetery, a Roman enclosure ditch, medieval and later drainage features and wells and a World War II aircraft crash site (Jonathan Sygrave, pers comm). But these low-lying wetlands were repeatedly abandoned owing to successive wetter conditions before large-scale flood defences were developed (Ed Wilson, pers comm).

These investigations at Managed Realignment Schemes are the most extensive excavations so far undertaken in coastal wetland areas. It seems probable that future works at such schemes will be one of the principal new sources of new information on historic coastal land-use. In this sense the physical loss of the sites is more than counter-balanced by increased understanding.

European collaboration

Given that modern national boundaries bear no relation at all to former early prehistoric coastlines, collaboration at a European level is plainly needed to help ensure consistencies of approach and recording. As part of Planarch 2 (2a: archaeological evaluation for wetlands), an EU Regional Development Fund Interreg IIIB project, Essex and Kent County Councils collaborated with the-then Dutch Rijksdienst voor het Oudheidkungig Bodermonderzoek (ROB), the Vlaams Institut voor het Onroerend Erfgoed (VIOE) and the Belgian Centrale Archaeologische Inventaris (CAI) (NK7, 8). Fieldwork took place in North Kent in June–July 2004, providing an introduction to recording methodology and means of access to sites. Colleagues from the continent were unfamiliar with the island-and-mudflat landscape of the Medway, since such land-scapes have been almost fully reclaimed in Belgium and the Netherlands. They were also accustomed to full retrieval of artefacts during survey, whereas in England these might just be noted on site, roughly dated and then left. GPS survey was likewise unfamiliar, as was downloading to the HER. Erosion and SMP aspects were largely unfamiliar on their intensely managed coastlines. Practical aspects, including health and safety, were discussed. Plainly this work did not add much information, but it did accustom archaeologists from Belgium and Holland to conditions and sites in the Thames Estuary.

In Essex meetings took place in May 2005. Practical fieldwork was focused on The Stumble in the Blackwater Estuary. Visits were also made to Managed Realignment Schemes in the county, including Abbots Hall Farm. Managed Realignment in Belgium and Holland would

Figure 6.20
Blakeney Chapel, Norfolk.
A 6th-century bracteate
brooch, either a casual loss
or derived from a burial.
(Reproduced by permission
of Historic Environment
Service, Norfolk County
Council)

rarely be undertaken, because of the low topography of land behind.

There has also been collaboration between researchers on both sides of the North Sea, EH and the Dutch heritage service, Rijksdienst voor het Cultureel Erfgoed, in developing a North Sea Prehistory and Management Framework (Peeters *et al* 2009). On a continent-wide basis, a four-year research network (SPLASHCOS) has been funded by the European Commission under its COST program (Cooperation in Science and Technology) from 2009 to 2013. Its aim is to 'bring together archaeologists, marine geophysicists, environmental scientists, heritage agencies, and commercial and industrial organizations interested in researching, managing and preserving the archives of archaeological and palaeoclimatic information locked up on the drowned prehistoric landscapes of the European continental shelf, and to disseminate that knowledge to a wider public'.

Collaboration with universities

Some university archaeological departments and researchers focus their attention on overseas sites. Fortunately several do not. Examples include the work of Professors David Sear (University of Southampton) and Martin Bell (University of Reading). Professor Sear has brought the use of a range of sonic techniques to the investigation of the sub-merged medieval town of Dunwich, providing extensive information on the surviving remains of the town, besides developing a methodology that will be applicable to other sites in comparable situations (Sear *et al* 2011). Professor Martin Bell at Reading, some of whose work has been referred to above (Bell 1990; 2007; 2013), has undertaken extensive research and fieldwork, especially in the Severn Estuary, defining the intertidal archaeology of the region and relating it to adjacent dry land archaeology. His work has also underpinned that of the Severn Estuary Research Committee, fostering other investigations.

In many parts of the country, university-based research has made a fundamental contribution to our knowledge of coastal and intertidal sites. Collaboration with universities also provides the prospect of Research Council funding to address key research questions, besides the investigation of threatened sites, as part of undergraduate training or PhD projects.

Given reduced funding for English Heritage, this opens new possibilities for investigation.

Monitoring

A report from any intertidal survey represents a set of observations at a given time. However, in dynamic coastal environments some sites rapidly erode while others are obscured by shifting sand or mud, or by FCERM structures. Concurrently, erosion may expose previously unknown sites. In the Severn Estuary subsequent visits to intertidal fish traps showed that some had been destroyed since previous recording, while new structures were revealed (S4). It follows that long-term monitoring is needed to produce something approaching a full record.

Following an initial survey in the 1980s (Wilkinson and Murphy 1995), key sites in Essex were re-examined over a three-year period from 2001–4 (E4). Reinspection in the early 2000s showed that midden-type and Red Hill deposits at Leigh Beck, Canvey Island, had virtually been lost to erosion. Some 50m of peat deposits had been lost from the western end of a peat outcrop and submerged forest at Purfleet, alongside masking of much of the site by concrete blocks placed to reinforce the sea wall. Erosion of the palaeosol and overlying peat sequence at Fenn Creek in the Crouch was less severe: there had been a retreat of the river bank of some 5m in 19 years, though this did not affect the overall exposure. At Alresford on the Colne an entirely new set of wooden structures was recorded, bearing no relation to the medieval structures previously planned. At Jaywick the prehistoric surface was invisible, masked by sand that had been emplaced following a major FCERM project involving granite breakwaters and recharge during the late 1990s, while at Clacton the beach is littered with boulders eroded from defences with intervening sand, all obscuring the palaeosol. At the mudflat of The Stumble in the Blackwater there was new exposure of the Neolithic palaeosol extending it by some 40m, while a newly exposed area of palaeosol, some 100 × 20m, was recorded close to the causeway and new wooden structures were observed eroding immediately to the south of the salt-marsh. At Collins Creek new exposures of wood were observed, while previously placed markers showed vertical erosion of some 0.25m. At Rolls Farm the original Neolithic site could not be relocated, but a new finds scatter to the west was observed. By contrast, erosion of mudflats had exposed six sections of wooden trackway, mostly not previously recorded.

On the Isle of Wight monitoring between 1990 and 1994 at Young's Slipway at Fishbourne, using direct measurement to plot a digital terrain model, showed a loss of 2,388 cubic metres of sediment from the foreshore, 57 per cent of which was lost in the first two years (Loader *et al* 2002, 29–30). The survival of sites from the Neolithic onwards in the vicinity implies that this rate of erosion in Wootton Creek is recent, partly related to ship wash, wave action and the draw-down of material into a dredged navigable channel. In addition, extensive damage is caused to intertidal surfaces by oyster dredging between Binstead Hard and Quarr, producing conspicuous gouging and pulling vertical posts from their vertical positions (Isle of Wight County Archaeology and Historic Environment Service 2005).

Monitoring of the promontory of Hengistbury Head has shown that upper Palaeolithic and Mesolithic sites are threatened by long-term erosion: map regression shows loss of up to 80m between 1885 and 2011. Recent work using LiDAR and Total Station Survey has defined the present situation and provides a basis for planning future recording (Cole *et al* 2012).

These few observations give some impression of the types of change that may occur, on balance representing losses of sites and information. However, in the longer term regular monitoring by professional archaeologists is unlikely to be practicable or affordable widely, except perhaps for sites of national significance. Involvement of avocational groups is the obvious solution.

Who owns the coastal historic environment?

Who *owns* it indeed? In a literal and legislative sense most sites below high-tide level are the property of the Crown Estate, but here we will consider who *feels* that they own it. This conflicted issue is illustrated here with reference to the excavation and recording of one site: the Bronze Age structure at Holme-next-the Sea, Norfolk, which became known as 'Seahenge' (Brennand and Taylor 2003). The site was first

'discovered' by a local volunteer archaeologist in the mid-1990s, and reported to Norfolk Landscape Archaeology, though local residents at Holme had known it was there for decades and said it had not changed over that time. It is certainly true that it would have *looked* the same for decades, but a vertical wooden post driven down into intertidal sediments will retain its surface appearance on the beach as it degrades, right down to its base. In fact, there was good evidence that this coast is highly dynamic, subject to dune recession and beach lowering, and that exposed surfaces of timbers were being degraded by marine timber-boring organisms. The site was on a trajectory to destruction by erosion. It was an unusual prehistoric monument that required detailed recording before it was eroded away.

A press release about the site had an unexpected outcome. It captured the attention of the media and this spread literally worldwide. Shortly, crowds of people arrived to see it and the small car-park at the Royal Society for Protection of Birds (RSPB) Reserve became full of media vans with satellite aerials, from many nations. It soon became necessary to control access around the excavation by means of temporary fencing, to provide pathways so as to avoid more extensive damage from trampling. The RSPB wanted an end to this as soon as possible, for the visitors were disturbing birds and damaging their feeding ground. They wanted the site excavated and removed. On the other hand there were local people who considered the site to be *theirs*, and wanted it to stay where it was. Soon after pagan groups intervened, claiming it as a sacred site and *their* own. Meanwhile the excavation put the team under severe stress: it was very hard physical labour for the excavation team, and at awkward hours, to fit with tides.

Eventually the outer timber circle had been excavated and removed and all that was left to do was to lift the central trunk. After a previous altercation when pagans had 'sat-in' on the monument (Fig 6.21), and David Miles, then Chief Archaeologist at English Heritage, talked with them, an injunction to exclude them had been put in force. The tackle was placed around the trunk and it began to be raised. Just then there was a sudden shout, and a young woman bounded towards the trunk. Dr Bill Boismier, aided by a policeman, felled her with a rugby tackle before she reached it – fortunately for her, for she would probably have been injured or killed had she made it. Some in the crowd shouted 'Shame on you!' and 'Police brutality!' Amidst all this, I was the first into the 1.5m deep pit left from removal of the trunk. I slithered rapidly down the wet clay side of the pit, aware that the tide was turning, and that the time left for observation and recording was extremely limited. We had no idea at all what might be in there for there was no precedent for this site. Looking around rapidly, and scraping judiciously with my trowel, I was relieved to find that there was *nothing* there beyond an extension of the plaited rope of stems of *Lonicera* (honeysuckle) used to drag the trunk into position (Fig 6.22). Maisie Taylor and I cleaned, recorded and collected the rope. Finally, offerings from the pagans – tears and flowers – were placed in the pit before all was submerged by the rising tide.

The intense feelings that this excavation aroused were a shock for which we were unprepared. We had been accustomed to the majority of archaeological interventions being unknown to most people, apart from an occasional 'open day'. The coast is a public place and so here we were exposed to the full light of controversy. The excavation at Holme raised very sharply the notion of ownership of the heritage and how best to manage it. The whole process rather smacked of the 'Decide. Announce. Defend.' approach then adopted by many public organisations when initiating any new project. There had been public meetings about the future of the site, but they had been confrontational, and largely negative in outcome. Longer-term consultation, discussion and negotiation, in which opinions and views could be debated calmly, and a consensus reached, were then not on the agenda, largely because of threats to the natural habitat.

This led to lower-key interventions at the site from 2003–2008, monitoring the erosion of other timber structures, repeatedly recording them as new elements became visible and removing individual timber components for detailed recording and scientific dating (Ames and Robertson 2009). This did not attract media attention and indeed few people were aware that archaeological work was still continuing. From a purely professional archaeological perspective this was successful: almost as much information was obtained as would have come from full excavation. It was far cheaper than excavation, and the work was unimpeded by objectors. However,

this approach could be viewed as bordering on the clandestine to achieve a specific archaeological end, and it did not encourage public participation or volunteers.

Volunteer participation has a long and respected history in English archaeology. There is nothing new about the involvement of volunteers in archaeological recording. A bronze rapier was recovered during the construction of the Royal Edward Dock at Avonmouth in 1903 from a depth of some 50 feet, apparently in channel sediments (Brett 1996). This find reflects an early phase of archaeological work alongside development,

Figure 6.21
Holme-next-the-Sea, Norfolk. Pagans 'sit-in' in the monument in an attempt to stop removal of timber components.

Figure 6.22
Holme-next-the-Sea, Norfolk. A rope of twisted honeysuckle stems used to drag the central tree into position in 2049 BC.

fostered by the high profile of local archaeological societies in Bristol at the time: several elected members of Bristol Corporation were active in archaeological research.

This is not to say that excavations undertaken by volunteer groups were always exemplary, and many sites were never published fully or at all. However, in the mid- to later 20th century, at a time when professional archaeologists were few, volunteers made an essential contribution, often working alongside professionals. This was the route through which many modern professional archaeologists, including the writer, entered the discipline. Collaboration between professional museum-based archaeologists and commercial developers began to change the ambience towards the end of the century. Local Authority archaeological services had been established from the 1970s onwards. The publication of 'Planning Policy Guidance Note 16: Archaeology and Planning' (PPG 16) in 1990 by the then-Department of the Environment led directly to the widespread establishment of commercial archaeological contractors and the professionalisation of English archaeology in general. Contractual obligations, tight deadlines and Health and Safety considerations made the new contracting archaeological companies reluctant to engage with 'amateurs', apart from some organisations that had been established as charitable trusts. As a result, volunteer involvement became generally more marginal, a trend consistently regretted by some writers. Nevertheless, an active archaeological voluntary sector has persisted and, in places, flourished (Fig 6.23).

In the light of the new political climate the volunteer tradition needs revisiting, resuscitation where necessary, and support. This is nowhere more needed, nor easier, than at the coast. The EH RCZAS field programme has recorded heritage assets visible during the individual surveys over a very short time span, often following only a single visit per site; but coastal change will mean that new, previously unknown, sites will become exposed and those recorded earlier will be eroded away. Due to the unpredictability of future funding, EH has rarely been able to fund long-term centrally funded monitoring programmes. This applies even more now. Encouraging and supporting local groups to monitor sites in their own areas is the obvious solution, not least because they are familiar with *their* places, and are

able to respond immediately to rapid changes, for example after storms. The very *visibility* of eroding intertidal archaeological sites makes them ideally suited for recording by volunteers, since the costs of machine stripping and the infrastructural elements of a modern terrestrial excavation are obviated.

This will, however, not be a cheap option. The inadequacies of some local groups in the last century were attributable to a lack of technical training and professional support, and to the lack of a mechanism supporting the archiving or publication of their results. These all need to be provided. Moreover, there are significant implications in terms of Health and Safety. Nevertheless a start has been made.

Involvement of volunteers has been undertaken as part of the New Forest RCZAS, for both intertidal and offshore survey (NF2, 21–6). This involved two training days, held at Brockenhurst and Lyndhurst, providing background information on coastal archaeology, the archaeology of the New Forest, and an introduction to identification of artefacts and sites, and surveying and recording (Fig 6.24). Volunteers participated both in archaeological survey and at the evaluation of the Creek Cottage saltern (Fig 6.25), where work took place as part of the festival of British Archaeology in July 2010. Finds were mainly of post-medieval date, although medieval pottery came from underlying silts and other contexts, helpfully suggesting an earlier origin for the site. Foundations of a coal store and probable foundations for the (now reduced) boiling house were defined (NF2, appendix 3). On the south coast, the eroding Roman villa and pre-existing Iron Age occupation on a chalk cliff at Folkestone, Kent, is eroding and is currently under excavation by the Hendon and District Archaeological Society, led by the Canterbury Archaeological Trust (Selkirk 2012). On the Isles of Scilly publicity about the Lyonesse Project led to local people volunteering during survey and information on palaeo-environmental deposits (Cornwall Council 2012, 195). Elsewhere, excavation of coastal sites has continued, for example by the Chichester and District Archaeological Society at a small Roman villa site at Warblington (Fig 6.26), and at sites such as Happisburgh, Norfolk (Fig 6.27). In the North-West, at Druridge Bay, Northumberland (following on from the RCZAS there), the Heritage Lottery Fund has provided £285,900 to support a

Figure 6.23
Collins Creek, Blackwater
Estuary, Essex. The late
Ron Hall, who alerted
professional archaeologists
to this and other sites, and
contributed enormously
to their recording.
(Image by Ellen Heppell
courtesy of and © Essex
County Council)

Figure 6.24
New Forest, Hampshire.
Field survey on the coast
adjacent to the Beaulieu
river. The volunteer survey
team, led by staff from
Wessex Archaeology,
examines wooden
structures.

Figure 6.25
Creek Cottage, Lymington, Hampshire. Volunteer workers excavate part of a saltern site.

Figure 6.26
Warblington, Hampshire. Volunteer workers from Chichester and District Archaeological Society excavate postholes of an aisled hall at a coastal Roman villa site.

project named 'Rescued from the Sea'. This is training and supporting a group of volunteers to record the site as it erodes, besides providing guided tours giving educational services for schools and opportunities for young offenders from HMP Northumberland to participate. The finds and records will be properly archived at the Great North Museum. This is plainly a very profitable way ahead.

Figure 6.27
Happisburgh, Norfolk. Dr Peter Robins, who found the first stratified hand-axe at this Palaeolithic site, waits to remove a barrow filled by the writer.

What next?

In an ideal world EH would provide funding for increasing flood resilience of listed buildings, would consider the feasibility of relocation of some structures, and would ensure archaeological recording of all significant sites under threat from erosion. In practice there was never sufficient funding to achieve all this, and this unfortunate fact is even more strongly the case at a time of economic stringency. EH's resources are limited and the application of them must be strategic. Only very rarely will funding for specific interventions be possible, and then only for sites of the highest national significance where other sources of funding cannot be obtained. The priorities we have identified are as follows:

- We will continue to support our programme of Rapid Coastal Zone Assessment surveys, as part of the National Heritage Protection Plan (NHPP);

- We will extend the GIS methodologies developed for our Coastal Estate Risk Assessment (CERA) to address inland flooding;

- We will aim to extend this approach to the wider historic environment following the completed second generation Shoreline Management Plans and consider how to integrate this work within our wider Heritage at Risk programme;

- We will address the designation of coastal heritage assets as part of our Heritage Protection Review;

- We will continue to work with Defra, the Environment Agency and others to develop the integration of historic environment considerations in flood and coastal erosion risk management; and

- We will explore the role the voluntary sector will play in responding to coastal change.

It is unfortunate that large-scale EH-funded excavation will rarely be affordable. However, new data will be obtained from professional investigations at Managed Realignment and other coastal development sites (such as expanding nuclear power stations and bridge construction) and from long-term monitoring of other locations, often by volunteers. Telling *them* that finding things is fun is pushing on an open door. We will have to cut our cloth according to our purse and manage the inevitable long-term losses so as to maximise the information we have on the development of the coastal historic environment from the Palaeolithic to today.

APPENDIX

Regional research priorities

The regional priorities identified here are largely drawn from Petts and Gerrard 2006; Manby *et al* 2003; Cooper 2006; Glazebrook 1997; Brown and Glazebrook 2000; Williams and Brown 1999; Nixon *et al* 2002; http://oxfordarchaeology.com; http://www.kent.gov.uk/leisure_and_culture/heritage/south_east_research_framework.aspx; Webster 2008; www.algao.org.uk/england/research_frameworks and Brennand 2007), supplemented by data from RCZAS reports and Fulford *et al* (1997). In part they reiterate national priorities discussed above but for each region there are specific points of interest. It is notable that in some regions the Regional Archaeological Research Frameworks have defined priorities that would apply equally well to other regions, but have just not been picked up.

North-East (Scottish border to Whitby)

Sites targeted for investigation during the field survey are listed in NE5, 32. These are all sites where coastal erosion is degrading significant historic assets. Priorities for future research in the North East, derived from Petts and Gerrard (2006) and Fulford *et al* (1997, 231–3), are as follows.

- Archaeology of the early post-glacial coastline: the place of hunter-gatherers in the North Sea littoral including submerged prehistoric landscapes.
- Multi-period assessment of the Tweed, Tyne, and Tees.
- More research is needed on salt production (including sites at Coatham Marsh, Tees).
- The early medieval coast. Research is needed on fisheries and possible intensification of deep-sea fishing from *c* AD 1000 with mapping of the early medieval coastline.
- Studies of sand-dune development, for example at Green Shiel, Lindisfarne. Dune systems in general require more investigation, involving surveys, regression studies and localised excavation, focusing attention on Druridge Bay and Low Hauxley.
- The impact of the Vikings.
- The fishing industry: more medieval fish-bone assemblages are needed.
- Transport and communication. More research is needed on post-medieval smaller ports and harbours, and on maritime infrastructure such as lighthouses and coastguard posts.

- Early coal industry and coal use. More research is required from the medieval period to the 19th century.
- Industrialisation. There is a need to identify medieval salt production sites using 'sleeching'. Direct boiling, generally supposed to have begun in Tyneside in the 15th century, needs more research. There should be further recording of coastal sites involved in the extraction of building stone, alum, coal, iron and jet as they erode.
- World War I military remains are seen as a priority for investigation.
- Shipbuilding. A desk-based assessment of all remains of shipbuilding in the North-East should be developed. More research on 19th–20th century steel-hulled wrecks is necessary.
- Wrecks. The maritime HER needs enhancement to include them.

Yorkshire, Humberside and Lincolnshire

The main sources referred to here are Manby *et al* (2003), Cooper (2006), Fulford *et al* (1997, 31–3) and YL5–8. Following the latter reports, further, more detailed, investigation of sites has been undertaken (Brigham and Fraser 2012a; 2012b; Brigham *et al* 2012). This study includes four main task headings:

- Historical audits and archaeological survey of selected harbour sites (Whitby, Scarborough and Bridlington, South Landing, Flamborough);
- Recording of identified World War I monuments and selected World War II monuments. These mainly consist of small square pillboxes in East Yorkshire, but a small number of monuments have also been identified in the historic county of Lincolnshire;
- Sampling and extent surveys of eroding meres and former land surfaces to include identification of timber species, plant macrofossil, diatom and pollen identification, and radiocarbon/AMS dating. The sites include the remains of eroding meres, relict land surfaces and infilled channels located on the foreshore in East Yorkshire and an area of relict land surface has also been revealed in Cleethorpes, north-east Lincolnshire; and

- Evidence for coastal industries, focusing on the Filey Brigg quarrying industry and the medieval–Early Modern site of Cayton Cliff Mill.

Other priorities for research include:

- Further research on Pleistocene and Holocene sediment units providing palaeoecological data is required. Deposit modelling will be desirable.

- Settlement and shoreline loss at Flamborough Head, Bridlington Bay, Holderness and along the Lincolnshire coast.

- Multi-period assessment of the Humber. Further survey to identify 2nd millennium BC boats and related timber structures should be undertaken.

- Further recording of coastal sites involved in the extraction of building stone, alum, coal, iron and jet should be done as they erode.

- Salt production. Monitoring of the Ingoldmells area should continue, although at present the sites are concealed by beach deposits. Salt mounds on the Lindsey coast require dating.

- The fishing industry. More medieval fish-bone assemblages are needed.

- Ports in Lincolnshire (Skegness, Spalding, Wainfleet, and Wrangle) are under-researched. Boston was the pre-eminent port in the region from the 11th century, and of national significance, but its archaeology is poorly understood.

The East of England and the Thames Estuary

Future research of this region is considered in the regional Research Frameworks for the East of England (Brown and Glazebrook 2000), Greater Thames Estuary (Williams and Brown 1999) and London (Nixon *et al* 2002), as well as Wilkinson and Murphy (1995) and Wilkinson *et al* (2012) for survey and excavation on the Essex coast. The relevant RCZAS reports are N1, S1, S3, S4, E1–5, and NK1–8. The first priority is initiation of RCZAS survey of the coast between the Medway and the North Foreland in Kent: funding problems led to this coast being omitted from earlier work. Fulford *et al* (1997) note several research objectives, which have been followed up, at least to some extent.

- For the Palaeolithic, collation of data and more research on the Thames palaeochannels and north Norfolk coast is needed. Deposit modelling is needed in general for all periods but especially the Palaeolithic, including evaluation of borehole data. This should be extended to include purposive surveys.

- Holocene sediment sequences requiring further work include coastal dunes, assessing in particular the potential of the Norfolk and Suffolk coasts. Further evaluation of potentially waterlogged Mesolithic sites is required, including that at Fenn Creek, South Woodham Ferrers, Essex.

- Additional survey/monitoring of the Essex coast and Medway coasts has largely been achieved through the RCZAS, but the recognition of the rarity of Neolithic sites such as The Stumble, Blackwater estuary, necessitates longer-term monitoring and research.

- London's hydrology and river systems need further attention, looking at the relationship between landscape, river and settlement. The prehistoric metalwork sequence from the Thames needs more attention. The role of catastrophe in the development of London is a fruitful line of approach.

- There has been a lack of attention to the Roman coast (*see* Murphy 2005). Similar studies are needed elsewhere. There should be more retrieval of fish bones from excavations, systematic investigation of possible port sites, and more consideration of the later use of saltern mounds. Offshore, shipwreck survey is needed (cf work at Pudding Pan and Pan Sands north of Whitstable).

- Settlement and shoreline loss are highlighted by Fulford *et al* (1997), including assessment of Roman find scatters and medieval records for Norfolk (Cromer (Shipden), Eccles and Suffolk (Covehithe, Dunwich, Walton Castle)). As noted *above*, some of these sites have received further investigation, but additional work is needed.

- Land reclamation all round the coast is understood in a fragmentary way archaeologically and by inference from documentary sources, but critical studies of significant areas are wanted.

- For the Anglo-Saxon period, definition of areas where palaeoenvironmental evidence may be expected to survive is needed. Full publication of excavations at the *wic* at Ipswich is necessary.

- Defence and defence-related sites require more study, including Landguard and Tilbury Forts, and Chatham Royal Naval Dockyard, where more attention to remodelling is needed.

- Industrial archaeology needs more attention, including specific detailed survey of barge and shipbuilding yards.

- Post-medieval canals, rivers and ports need more attention everywhere. London's docklands and waterways need further study to increase understanding of Roman London's role as a port and centre of trade and how this changed through time, as well as identifying materially (through the archaeological record) how London became a redistribution centre for the western world. Medieval ports elsewhere, for example King's Lynn and Wisbech, also need more research, though this is likely to be achieved mainly through excavations undertaken as part of development control.

- Presenting data on past coastal changes to the public may help to make modern coastal management schemes more acceptable.

South coast, including the Solent

Phase 1 of the RCZAS (desk-based survey of the south coast) has been completed. Other RCZAS reports are IoW, and NF1–3. The main priority for the region now is initiation of Phase 2 (Field Survey) when NHPP funding for Activity 3A2 permits. The Solent–Thames Archaeological Research Framework covers Hampshire and the Isle of Wight and the South-East Archaeological Framework Kent and Sussex (http://oxfordarchaeology. com; and http://www.kent.gov.uk/leisure_and_culture/ heritage/south_east_research_framework.aspx). Both include some research priorities relevant to the coast. Poole Harbour and Selsey Bill were highlighted by Fulford *et al* (1997) as priorities for study; the former has been studied by the Poole Harbour Heritage Project (Dyer and Darvill 2010), while current excavations at Medmerry will add understanding to the latter.

- For the Palaeolithic, better definition of the Sussex raised marine deposits is needed, extending the Boxgrove Raised Beach Mapping across the county. More reliable dating is required, for example, of the Aldingbourne Raised Beach, which has been attributed to Marine Isotope Stage (MIS) 7 or 11. Raised Beaches in Sussex need correlation with fluvial terrace sequences. Do raised beach deposits occur between Cowes and Bembridge? Further investigation of the Priory Bay and Bembridge Raised Beach sites is a necessity.

- Dating the first isolation of the Isle of Wight from the mainland.

- Further survey of coastal and submerged upper Palaeolithic and Mesolithic archaeology in the Solent is needed, especially since these sites are likely to produce palaeoenvironmental data, as at Bouldnor Cliff. Predictive modelling of deposits is needed to guide field survey.

- Knowledge of the extent, function and dating of Neolithic and Bronze Age timber structures in Langstone Harbour and at Wootton-Quarr needs to be extended.

- The exploitation of fish and shellfish at Roman sites, including evidence for aquaculture needs further research. This will require further sampling at excavated sites.

- Evidence for the maritime use of the Solent in the Roman period needs attention. Better definition of waterlogged coastal harbours, jetties, and boats is required and evidence for development of ports, including Clausentum (Bitterne), demands investigation.

- Coastal change and land reclamation in the early medieval period needs further study, as has been done at Romney Marsh and Pevensey Levels. Evidence for the impacts of the Little Ice Age on the coast is not well understood.

- The early Anglo-Saxon development of ports and *wics*, including Fordwich, Sandwich and Pagham Harbour, needs study. The environs of coastal 'productive sites' need evaluation to determine means of access from the sea.

- Domesday and later documentary sources record watermills, tide-mills, salterns and fisheries (fish ponds, fish traps), but archaeological evidence for early development remains slight (IoW, NF1–3).

- Stone quarrying in the Solent, Portland Bill and other locations needs more investigation.

- The development of ports in general and of coastal inlets on the Isle of Wight needs more attention.

- Later medieval coastal military defences, especially those related to known political events are not well understood. Coastal defences of all periods, and related structures need more detailed survey. Offshore World War II aircraft wrecks need special attention.

- Ship- and boat-building was, in the pre-modern era, a significant part of the south coast's economy. More detailed studies are needed, comparable to that undertaken at Buckler's Hard.

- Coastal leisure and recreational activities have become economically significant since the 19th century. More research is needed, building on Brodie and Winter (2007).

South-West (Devon and Cornwall) and Isles of Scilly, including the Severn Estuary

The RCZAS of the Isles of Scilly has been completed, (IoS), but survey of the remainder of the region requires completion. Relevant RCZAS reports for the Severn Estuary are S1–4. To a large extent the latter survey has met the research priorities defined by Fulford *et al* (1997, 232–3). NMP recording for Cornwall has been completed and an NMP project to cover the AONB coast of North Devon is underway. As part of NHPP Activity 3A2, Phase 1 of the RCZAS for south-west England (south coast) has been initiated. A Research Agenda for the South-West (edited by C Webster) is available on somerset.gov.uk/ archives/hes/downloads/swarf_15.pdf. Research aims from that publication relating specifically to coasts are included here.

- Better understanding of coastline change is needed for all periods, but especially for the Palaeolithic and Mesolithic.

- There is little evidence for coastal middens or complex mortuary rites in the Mesolithic: have they been lost to sea-level rise? Sites such as Westward Ho! need linking to the terrestrial record. Is there evidence for management of coastal resources such as reed-beds at this period?

- More detailed study of submerged forests and coastal peats is needed to provide proxy data for Mesolithic and Neolithic environments. At all periods, the use of salt-marshes to provide plant resources needs investigation.

- Development of deposit models, linked to prospection, is necessary for all periods but particularly the early prehistoric periods.

- Further use of LiDAR for better definition of surface infilled palaeochannels is needed.

- More survey work in river valleys and coastal plains is required.

- On Scilly further survey of field walls and associated features in the intertidal zone is wanted.

- Coastal sites such as the Isles of Scilly and Gwithian provide opportunities for micro-morphological studies of soils to consider soil management.

- Coastal dunes: the dunes of North Cornwall need special attention.

- Estuaries: there should be multi-period assessments of the Camel, Exe, Tamar and Fal. Monuments and settlements: assessments of Yelland (Taw–Torridge), Devon with Mesolithic to Bronze Age lithic scatters and intertidal stone row are necessary, but survey of these areas will be achieved as part of the SW RCZAS.

- Roman-period salt production has been investigated extensively, but little is known of salterns of later date.

- Production of alum and copperas in the region has been neglected.

- Ports, and putative ports, of all periods demand more investigation.

- Major fixed military defences have been studied, but more work is needed on earthen structures, from the Civil War onwards. Coastal artillery defences need more scrutiny. For the latest period, more attention should be paid to features related to the major defences, in terms of adaptation for 'total war' in the 20th century.

North-West

In 1997 Fulford and his colleagues had little to recommend for the coast of this region. While not entirely *terra incognita* it was certainly poorly investigated. Their recommendations included:

- Survey of coastal dunes – Cumbria and Lancashire.

- Surveys of estuaries – Dee, Mersey, Ribble and Wyre.

- Settlement and shoreline loss – Wirral (eg Meols) and Lancashire.

The NW RCZAS has, to a large extent, met these aims (NW1–4). In addition, the North-West Archaeological Research Framework Volume 2: Research Agenda and Strategy (Brennand 2007) provides additional objectives. Priorities identified in these documents include:

- There is no provision for emergency rescue excavation of maritime sites and no contractor in the region with specialist experience. There is no curation of resource below Mean Sea Level. A maritime HER is needed, but resources would be required to develop it. An assessment of maritime finds and an evaluation of possible wreck sites are needed.

- Relative Sea Level and coastal change are only understood at a broad scale. Most RSL research was undertaken some time ago and has little direct linkage with archaeology.

- Offshore Palaeolithic and early Mesolithic submerged sites are unknown but should be detectable from survey, followed by targeted excavation.

- There is a need for continued recording of submerged forests and prehistoric human and animal footprints (notably at Formby). Extensive excavations of Mesolithic sites on West Cumbrian coast such as Eskmeals need publication.

- For all periods there is potential for survival of buried land surfaces beneath dune systems.

- The inland salt industry is relatively well investigated but more work is needed on coasts.

- Further investigation of the eroding Swarthy Hill hillfort is required.

- A programme of survey is needed to examine potential Roman fort sites, especially on the Cumbrian coast. Although the trading site at Meols was largely destroyed by erosion in the 19th century there is still potential for associated archaeology to survive behind the modern sea wall. Roman ports in the region are poorly understood. For the Roman period, survey should target estuaries and major estuaries, especially the Dee, to determine palaeogeography and navigability.

- For the early medieval period priorities are similar to those for the Roman period: further work is needed at Meols and studies on the palaeogeography and navigability of the main estuaries and rivers is needed. No major early medieval settlements are known between Meols and Whithorn in Galloway. Survey is needed to detect potential trading sites: coin and artefact finds from coastal areas could be indicative.

- Systematic survey of coast and hinterland is needed, paying especial attention to fishing in the medieval period. Further work is needed on the role of monastic communities in coastal salt production and other industries. Survey to detect medieval wreck sites is needed. Eroding sites such as the Aldingham motte-and-bailey castle and Cockersand Abbey are eroding and demand further investigation.

- Landscape archaeology projects are needed to investigate effects of the Little Ice Age and coastal change. Survey for of evidence for the post-medieval coastal salt industry and fish traps is required on the Solway, and in Cumbria, Morecambe Bay and Fylde coast. Little is known archaeologically about the region's ports. Targeted surveys, for example of Chester and Liverpool, are needed to see why one took over from the other.

- For the Modern and Industrial periods, survey is needed of wreck sites in Liverpool Bay, intertidal structures in Morecambe Bay, and at Fleetwood, one of the country's premier fishing ports in the 19th and 20th centuries. Early facilities in Liverpool Docks are currently being investigated, but targeted investigation of warehousing is needed. A regional study of late 18th- and 19th-century military coastal defences is required. They are poorly represented in SMR/HERs, even at Liverpool and Whitehaven. Further field investigation of World War II anti-invasion defences is necessary. The physical evidence for the development of coastal resorts requires further scrutiny.

BIBLIOGRAPHY

Albone, J, Massey, S and Tremlett, S 2007 *The Archaeology of Norfolk's Coastal Zone. Results of the National Mapping Programme. English Heritage Project 2913.* Gressenhall: Norfolk Landscape Archaeology/English Heritage

Allen, J R L and Fulford, M G 1990 'Romano-British and later reclamation on the Severn salt marshes at Elmore, Gloucestershire'. *Transactions of the Bristol and Gloucester Archaeological Society* **108**, 17–32.

Allen, M J and Gardiner, J 2000 *Our Changing Coast. A Survey of the Intertidal Archaeology of Langstone Harbour, Hampshire.* CBA Research Report **124**. York: Council for British Archaeology

Allen, T, Cotterill, M and Pike, G 2001 'Copperas: an account of the Whitstable copperas works and the first major chemical works in England'. *Industrial Archaeology* **23**, 93–112

Ames, J and Robertson, D 2009 *The Archaeology of Holme Beach: an archaeological monitoring survey of the intertidal zone, 2003–8.* NAU Archaeology Report **1444**. Norwich: NAU

Ashwin, T 2005 'Norfolk's first farmers: Early Neolithic Norfolk (c 4000–3000 BC); and Late Neolithic and Early Bronze Age Norfolk (c 3000–1700 BC)', *in* Ashwin, T and Davidson, A *An Historical Atlas of Norfolk* (3rd edn), 17–19. Norwich: Norfolk Museums Service

Bayliss, A 1998 'Some thoughts on using scientific dating in English archaeology and building analysis for the next decade', *in* Bayley, J (ed) *Science in Archaeology. An agenda for the future*, 95–108. London: English Heritage

Bayliss, A, Groves, C, McCormac, G, Baillie, M, Brown, D and Brennand, M 1999 'Precise dating of the Norfolk timber circle'. *Nature* **402**, 479

Bell, M G 1990 *Brean Down Excavations 1983–1987.* English Heritage Archaeological Report **15**. London: English Heritage

Bell, M G 2007 *Prehistoric Coastal Communities: the Mesolithic in Western Britain.* CBA Research Report **149**. York: Council for British Archaeology

Bell, M 2013 *The Bronze Age in the Severn Estuary.* CBA Research Report **172**. York: Council for British Archaeology

Bell, M G and Brown, B 2009 *Southern Regional Review of Geoarchaeology: Windblown Deposits.* Research Department Report 005/2009. Swindon: English Heritage

Bell, M and Warren, G 2013 'The Mesolithic', *in* Ransley, J and Sturt, F 2013, 30–49

Bicket, A 2011 *Submerged Prehistory: Marine ALSF Research in Context Marine ALSF Science Monograph Series No. 5.* (ed Gardner, J). Salisbury: Wessex Archaeology

Biddulph, E, Foreman, S, Stafford, E, Stansbie, D and Nicholson, R 2012 'London Gateway. Iron Age and Roman salt making in the Thames Estuary' *Oxford Archaeology Monograph* **18**. Oxford: Oxford Archaeology

Birks, C 2003 *Report on the Archaeological Evaluation at Blakeney Freshes, Cley next the Sea, Norfolk. NAU Report* **808**. Norwich: NAU

Bowden, M and Brodie, A 2011. *Defending Scilly.* Swindon: English Heritage

Boylan, P J 1967 'The Pleistocene mammalia of the Sewerby–Hessle buried cliff, East Yorkshire'. *Proc Yorks Geol Soc* **36**, 115–25

Bradley, R, Chowne, P, Cleal, R M J, Healey, F and Kinnes, I 1993 *Excavations on Redgate Hill, Hunstanton, Norfolk, and at Tattersall Thorpe, Lincolnshire.* East Anglian Archaeology Report No. 5. Field Archaeology Division, Norfolk Museums Service/Heritage Trust of Lincolnshire. Gressenhall/Sleaford

Brennand, M and Taylor, M 2003 'The survey and excavation of a Bronze Age timber circle at Holme-next-the-Sea, Norfolk, 1998–9'. *Proceedings of the Prehistoric Society* **69**, 1–84

Brennand, M 2007 (ed) *North West Archaeological Research Framework, Volume 2 Research Agenda and Strategy.* English Heritage, The Council for British Archaeology and ALGAO (http://www.liverpoolmuseums.org.uk/mol/archaeology/arf/)

Brett, J 1996 'Archaeology and the construction of the Royal Edward Dock, Avonmouth, 1902–1908'. *Archaeology in the Severn Estuary* **7**, 115–20. Exeter: SELRC

Brigham, T and Fraser, J 2012a Rapid Coastal Zone Assessment Survey. Yorkshire and Lincolnshire: Phase 3. Site Survey and Historical Summary. Flamborough Medieval Harbour, Flamborough, East Riding of Yorkshire. *Humber Field Archaeology Report* **416**. Kingston-upon-Hull: HFA

Brigham, T and Fraser, J 2012b Rapid Coastal Zone Assessment Survey. Yorkshire and Lincolnshire: Phase 3. Historical Audit. Bridlington Harbour, East Riding of Yorkshire. *Humber Field Archaeology Report* **417**. Kingston-upon-Hull: HFA

Brigham, T, Buglass, J and Jobling, D 2012 Rapid Coastal Zone Assessment Survey. Yorkshire and Lincolnshire: Phase 3. Field Survey and Assessment. Selected First and Second World War Monuments. North Yorkshire, Lincolnshire, East Riding of Yorkshire. *Humber Field Archaeology Report* **415**. Kingston-upon-Hull: HFA

Brodie, A and Winter, G 2007 *England's Seaside Resorts.* Swindon: English Heritage

Brown, N and Glazebrook, J (eds) 2000 *Research and Archaeology: a framework for the Eastern Counties. 2. Research Agenda and Strategy.* East Anglian Archaeology Occasional Paper **8**

Carver, M 1998 *Sutton Hoo. Burial Ground of Kings?* London: British Museum Press

Carver, M and Loveluck, C 2013 'Early Medieval, AD 400 to 1000', *in* Ransley, J and Sturt, F 2013, 113–37

Chadwick, A M and Catchpole, T 2010 'Casting the net wide: mapping and dating fish traps through the Severn Estuary Rapid Coastal Zone Assessment Survey'. *Archaeology in the Severn Estuary* **21**, 47–80

Cole, J, Dominic, B and Davies, W 2012 *Report on the Topographic Survey of Hengistbury Head*. Southampton: University of Southampton

Coles, B J 1998 'Doggerland: a speculative survey'. *Proceedings of the Prehistoric Society* **64**, 45–81

Cooper, N J (ed) 2006 *The Archaeology of the East Midlands. An Archaeological Resource Assessment and Research Agenda*. Leicester: Leicester Archaeological Monographs **13**

Cornwall Council 2012 'The Lyonesse project. A study of the evolution of the coastal and marine environment of the Isles of Scilly'. *Report No*. **2012R069**. Truro: Historic Environment, Cornwall Council

Cracknell, B 2005 *Outrageous Waves. Global Warming and Coastal Change in Britain Through 2000 Years*. Chichester: Phillimore

Cunliffe, B 1971 *Fishbourne Roman Palace*. Stroud: Tempus Publishing Ltd

Cunliffe, B 1975 *Excavations at Portchester Castle Volume* **1**. London: Society of Antiquaries

Cunliffe, B 1987 *Hengistbury Head, Dorset. Volume 1: The Prehistoric and Roman settlement 3500BC–AD500*. Oxford University Committee for Archaeology Monograph **13**. OUCA: Oxford

Cunliffe, B 1988 *Mount Batten, Plymouth*. Oxford Committee for Archaeology Monograph **26**. Oxford: Oxford University Press

Cunliffe, B 2010 *Fishbourne Roman Palace. A Guide to the Site*. Chichester: Sussex Archaeological Society

Davies, L 2011 *Hulk Assemblages: Assessing the National Context. Final Report. English Heritage Project No*. **5919**. London: Museum of London Archaeology

Defra 2002 *Futurecoast* [A CD-based dataset on coastal change in England and Wales]. London: Defra

Defra 2004 *Making Space for Water. Developing a new Government strategy for flood and coastal erosion risk management in England*. London: Defra

Defra 2006 *Shoreline Management Plan Guidance. Volume 1: Aims and Requirements and Volume 2: Procedure*. London: Defra

Defra 2008 *A Strategy for Promoting an Integrated Approach to the Management of Coastal Areas in England*. London: Defra

Defra 2010 *Adapting to Coastal Change: Developing a Policy Framework*. London: Defra

Defra 2012 *UK Climate Change Risk Assessment 2012: Evidence Report*. (Defra Project Code **GA02024**). Defra, Welsh Government, DOE Northern Ireland, The Scottish Government. http://www.defra. gov.uk/environment/climate/government/

Dellino-Musgrave, V and Ransley, J 2013 'Early Modern and Industrial', in Ransley, J and Sturt, F 2013, 164–85

De Loecker, D 2011 *Great Yarmouth Dredging Licence Area 240, Norfolk, United Kingdom. Preliminary report on the lithic artefacts*. Unpublished report: Universiteit Leiden.

Department for Culture Media and Sport 2010 *Scheduled Monuments. Identifying, protecting, conserving and investigating nationally important archaeological sites under the Ancient Monuments and Archaeological Areas Act 1979*. London: DCMS http://www.culture.gov.uk/images/publications/ ScheduledMonuments.pdf

Devoy, R J N 1979 'Flandrian sea-level changes and vegetational history in the Lower Thames Estuary'. *Phil. Trans. Roy Soc London* **B285**, 355–407

DeSilvey, C, Naylor, S and Sackett, C (eds) 2011 *Anticipatory History*. Exeter: Uniform Books

Dover Harbour Board 2007 *Planning for the Next Generation: Second Round Consultation (January 2007)*. Dover: DHB

Dyer, B and Darvill, T 2010 *The Book of Poole Harbour*. Wimborne Minster: Poole Harbour Heritage Project

Emu and University of Southampton 2009 *Outer Thames Estuary Regional Environmental Characterisation. MEPF 08/01*. Marine Aggregates Levy Sustainability Fund. See also www.alsf-mepf. org.uk

English Heritage 2003a *Where on earth are we? The Global Positioning System (GPS) in Archaeological Field Survey*. Swindon: English Heritage

English Heritage 2003b Coastal Defence and the Historic Environment. Swindon: English Heritage

English Heritage 2006 *Shoreline Management Plan Review and the Historic Environment: English Heritage Guidance*. Swindon: English Heritage

English Heritage 2007 *Geoarchaeology. Using Earth Sciences to Understand the Record*. Swindon: English Heritage

English Heritage 2008a *Geophysical Survey in Archaeological Field Evaluation*. Swindon: English Heritage

English Heritage 2008b *Micro-generation in the Historic Environment*. Swindon: English Heritage

English Heritage 2008c *Climate Change and the Historic Environment*. Swindon: English Heritage

English Heritage 2008d *Luminescence Dating. Guidelines on using luminescence dating in archaeology*. Swindon: English Heritage

English Heritage 2010a *Flooding and Historic Buildings*. Swindon: English Heritage

English Heritage 2010b *Micro Wind Generation and Traditional Buildings*. Swindon: English Heritage

English Heritage 2010c *Waterlogged Wood. Guidelines on the Recording, Sampling, Conservation and Curation of Waterlogged Wood*. Swindon: English Heritage

English Heritage 2011a *Environmental Archaeology. A Guide to the Theory and Practice of Methods, from Sampling and Recovery to Post-Excavation*. Swindon: English Heritage

English Heritage 2011b *Small Scale Solar Electric (Photovoltaics) Energy and Traditional Buildings*. Swindon: English Heritage

English Heritage 2012a *National Heritage Protection Plan 2011–15. Year-End Report May 2011–March 2012*. Swindon: English Heritage

English Heritage 2012b *Designation Scheduling Selection Guide. Maritime and Naval*. Swindon: English Heritage

English Heritage 2012c *Waterlogged Organic Artefacts. Guidelines on their recovery, analysis and conservation*. Swindon: English Heritage

English Heritage nd *Dendrochronolgy. Guidelines on Producing and Interpreting Dendrochronological Dates*. London: English Heritage

Environment Agency 2007 *The Historic Environment and Managed Coastal Re-alignment: A Four Stage Approach*. Internal document: Environment Agency

Environment Agency 2011 *The Extent of Saltmarsh in England and Wales: 2006–2009*. Bristol: Environment Agency

Essex Biodiversity Project 1999 *Essex Biodiversity Action Plan: A Wild Future for Essex*

Fawn, A J, Evans, K, McMaster, I and Davies G M R 1990. *The Red Hills of Essex*. Colchester: Colchester Archaeological Group

Flemming, N C (ed) 2004 *Submarine Prehistoric Archaeology of the North Sea. Research Priorities and Collaboration with Industry*. CBA Research Report **141**. York: Council for British Archaeology

Flett, Sir John Smith 1937 *The First Hundred Years of the Geological Survey of Great Britain*. London: HMSO

Foot, W 2006 *Beaches, Fields, Street and Hills. The Anti-Invasion Defences of England, 1940*. CBA Research Report **144**. York: English Heritage and Council for British Archaeology

Frere, S S 1967 *Britannia. A History of Roman Britain*. London: Routledge and Kegan Paul

Fulford, M, Champion, T and Long, A 1997 *England's Coastal Heritage. A Survey for English Heritage and the RCHME*. English Heritage Archaeological Report **15**. London: EH/RCHME

Gaffney, V, Thomson, K and Fitch, S 2007 *Mapping Doggerland. The Mesolithic Landscapes of the Southern North Sea*. Oxford: Archaeopress

Galloway, A 2009 'Storm flooding, coastal defence and land use around the Thames estuary and tidal river *c*. 1250–1450'. *Journal of Medieval History* 35, 171–88

Gibbard, P L and Cohen, K M 2008 'Global chronostratigraphical correlation table for the last 2.7 million years'. *Episodes* 31, 243–7

Glazebrook, J (ed) 1997 *Research and Archaeology: a Framework for the Eastern Counties, 1. Resource Assessment*. East Anglian Archaeology Occasional Paper **3**

Gonzalez, S and Cowell, R W 2007 'Neolithic coastal archaeology and environment around Liverpool Bay', *in* Sidell, J and Haughey, F (eds) *Neolithic Archaeology in the Intertidal Zone*. Neolithic Studies Group Seminar Papers **8**, 11–25. Oxford: Oxbow Books

Greatorex, C 2003 'Living in the margins? The Late Bronze Age landscape of Willingdon Levels', *in* Rudling, D (ed) *The Archaeology of Sussex to AD 2000*, 89–100. Brighton: University of Sussex

Grieve, M 1959 *The Great Tide*. Chelmsford: Essex County Council

Griffiths, D, Philpott, R A and Egan, G 2007 *Meols. The Archaeology of the North Wirrall Coast. Discoveries and Observations in the 19th and 20th Centuries, with a Catalogue of Collections*. Oxford School of Archaeology Monograph **68**. Oxford: Institute of Archaeology

Hamel, A T 2011 'Wrecks at sea: ALSF research in context'. *Marine ALSF Science Monograph Series* No. **6**. Gardner, J (ed) **MEPF10/P152**. Salisbury: Wessex Archaeology

Hazell, Z J 2008 'Offshore and intertidal peat deposits, England – a resource assessment and development of a data base'. *Environmental Archaeology* 13, 101–10

Heppell, E 2006 'The Stumble, Essex', *in* Dyson, L, Heppell, E, Johnson, C and Pieters, M *Archaeological Evaluation of Wetlands in the Planarch Area of North West Europe*. Planarch 2 Interreg Project, Action 2A Report, 23–38. Maidstone: Kent County Council

Hijma, M P and Cohen, K M 2010 'Timing and magnitude of the sea-level jump preluding the 8200 yr event'. *Geology* **38**, 275–8

Hill, J D and Willis, S 2013 'Middle Bronze Age to the end of the Pre-Roman Iron Age, *c* 1500 BC to AD 50', *in* Ransley, J and Sturt, F 2013, 75–92

Houghton, Sir J 2009 *Global Warming. The Complete Briefing* (4th edition). London: Cambridge University Press

Hunt, A 2011 *English Heritage Coastal Estate Risk Assessment*. Research Department Report Series No. 68-2011. Swindon: English Heritage

Ings, M and Murphy, F 2011 *The Lost Lands of our Ancestors. Exploring the Submerged Landscapes of Prehistoric Wales*. Llandeilo: Dyfed Archaeological Trust

Isle of Wight County Archaeology and Historic Environment Service 2005 *Wootton–Quarr Beach Monitoring. Final Report February 2005*. Ventnor: Isle of Wight Council

James, J W C, Pearce, B, Coggan, A, Leivers, M, Clark, R W E, Hill, J F, Arnott, S H, Bateson, L, De-Burgh Thomas and Baggaley, P A 2011 *The MALSF Synthesis Study in the Central and Eastern English Channel*. Marine Aggregates Levy Sustainability Fund (MALSF). British Geological Survey Open Report 0R/11/01 **MEPF 09/P92**. See also www.alsf-mepf.org.uk

Jecock, M, Dunn, C, Carter, A and Clowes, M 2003 'The alum works and other industries at Kettleness, North Yorkshire: an archaeological and historical survey'. *Archaeological Investigation Report Series* **A1/24/2003**. Swindon: English Heritage.

Jenkins, G, Perry, M C and Prior, J 2009a *The Climate of the UK and Recent Trends*. Revised edition, May 2010. Exeter: Met Office Hadley Centre

Jenkins, G J, Murphy, J M, Sexton, D M H, Lowe, J A, Jones, P and Kilsby, C G 2009b *UK Climate Change Projections: Briefing Report*. Exeter: Met Office Hadley Centre

Jordan, M 2004 *Blakeney Freshes, Cley next the Sea, Norfolk. Archaeological Report*. Lindsay Archaeological Services Report **756**. Lincoln: LAS

Keene, L 1988 'Medieval salt working in Dorset'. *Proceedings of the Dorset Natural history and Archaeological Society* **109**, 25–8

Lambeck, K, Johnston, P, Smither, C, Fleming, K and Yokoyama, Y 1995 *Late Pleistocene and Holocene Sea-level Change*. http://rses.anu. edu.au/geodynamics/AnnRep/95/AR-Geod95.html

Lane, T and Morris, E 2001 (eds) *A Millennium of Salt-making: Prehistoric and Romano-British Salt Production in the Fenland*. Lincolnshire Archaeology and Heritage Reports Series No. **4**. Heckington: Heritage Trust of Lincolnshire

Lawrence, G F 1929 'Antiquities from the Middle Thames'. *Archaeological Journal* **XXXV**, 69–98

Lee, R 2005 *Blakeney Freshes, Cley-next-the-Sea, Norfolk. Archaeological Excavation NGR TG 0435 4525, Site Code 37793 CLY. MAP2 Assessment Report for the Environment Agency*. Lindsey Archaeological Services Report No. **817**. Lincoln: LAS

Loader, R D, Westmore, I and Tomalin, D 1997 *Time and Tide. An archaeological survey of the Wootton–Quarr Coast*. Ventnor: Isle of Wight Council

Lowe, J A, Howard, T P, Pardaens, A, Tinker, J, Holt, J, Wakelin, S, Milne, G, Leake, J, Wolf, J, Horsburgh, K, Reeder, T, Jenkins, G, Ridley, J, Dye, S and Bradley, S 2009 *UK Climate Change Projections Science Report: Marine and Coastal projections*. Exeter: Met Office Hadley Centre

Manby, T G, Moorhouse, S and Ottaway, P 2003 *The Archaeology of Yorkshire. An Assessment at the Beginning of the 21st Century*. Yorkshire Archaeological Society Occasional Paper No. **3**. Leeds

Marsden, P 2011 *The Levelling Sea. The Story of a Cornish Haven in the Age of Sail*. London: Harper Press

McAvoy, F 1995 Excavation at the Withow Gap, Skipsea (Holderness). *Humber Wetlands Survey 1st Annual Report:* Hull

McInnes, R 2008 *Coastal Risk Management – A Non-Technical Guide*. Ventnor: SCOPAC

McInnes, R 2011 *A Coastal Historical Resource Guide for England*. London: The Crown Estate

Megaw, J V S, Thomas, A C and Wailes, B 1961 'The Bronze Age settlement at Gwithian, Cornwall'. *Proceedings of the West Cornwall Field Club* **5**, 200–5

Miller, P, Mills, J and Bryan, P 2008 'Scanning for change: assessing coastal erosion at Whitby Abbey Headland'. *Research News* **9**, 4–7. Swindon: English Heritage

Milne, G, McKewan C and Goodburn, D 1998 *Nautical Archaeology on the Foreshore. Hulk Recording on the Medway*. Swindon: Royal Commission on the Historical Monuments of England.

Milner, N, Craig, O E, Bailey, G N, Pedersen, K and Andersen, S H 2004 'Something fishy in the Neolithic? A re-evaluation of stable isotope analysis of Mesolithic and Neolithic coastal populations'. *Antiquity* **78**, 9–22.

Momber, G, Tomalin, D, Scaife, R, Satchell, J and Gillespie, J 2011 *Mesolithic Occupation at Bouldnor Cliff and the Submerged Prehistoric Landscapes of the Solent*. CBA Research Report **164**. York: Council for British Archaeology

Murphy, P 1994 'The environmental evidence: introduction, mollusca, miscellaneous faunal remains, plant macrofossils, wood, summary and discussion', *in* Ayers, B S *Excavations at Fishergate, Norwich, 1985*. East Anglian Archaeology **68**, 34–61

Murphy, P 2005a 'Coastal change and human response', *in* Ashwin, T and Davidson, A (eds) *An Historical Atlas of Norfolk* (3rd edn), 6–7. Norwich: Norfolk Museums Service

Murphy, P 2005b 'Molluscs, plant macrofossils, environment and economy: a summary', *in* Crowson, A, Lane, T, Penn, K and Trimble, D 2005 *Anglo-Saxon Settlement on the Siltland of Eastern England*. Lincolnshire Archaeology and Heritage Reports Series No. **7**, 228–63. Heckington: Heritage Trust of Lincolnshire

Murphy, P 2007 'The submerged prehistoric landscapes of the southern North Sea: work in progress'. *Landscapes* **8 (1)**, 1–22

Murphy, P 2009 *The English Coast: a History and a Prospect*. London: Continuum UK

Murphy, P 2010 'The landscape and economy of the Anglo-Saxon coast', *in* Higham, N J and Ryan, M J (eds) *The Landscape Archaeology of Anglo-Saxon England*, 211–22. Woodbridge: The Boydell Press

Murphy, P and Green, F M L 2003 'Palaeogeography of the monument', *in* Brennand, M and Taylor, M 2003, 59–61

Nixon, T, McAdam, E, Tomber, R and Swain, H 2002 (eds) *A Research Framework for London Archaeology 2002*. London: Museum of London

Norfolk Archaeological Unit 2003 *An Archaeological Walk Over Survey at Holme Beach, Holme-next-the-Sea, Norfolk. Assessment Report and Updated Project Design*. Norwich: NAU

O'Sullivan, D and Young, R 1991 'The early medieval settlement at Green Shiel, Northumberland'. *Archaeologia Aeliana* 5th Series, **19**, 57–69

Oswald A, Ashbee, K, Porteous, K and Huntley, J 2006 *Dunstanburgh Castle, Northumberland; Archaeological, Architectural and Historical Investigations*. English Heritage Research Department Report **26/2006**. English Heritage: Swindon

Ottaway, P 2001 'Excavations of the site of the Roman Signal Station at Carr Naze, Filey, 1993–94'. *The Archaeological Journal* **157**, 148–82

Oxford Archaeology 2009 *London Gateway Compensation Site A. Interim Summary of Archaeological Results*. OA Job **4423**. Oxford: Oxford Archaeology

Palma, P 2009 *Swash Channel Wreck: Project Report for Environmental Scoping Study of in situ Stabilisation of the Wreck*. Poole: Bournemouth University

Parfitt, S A, Barendregt, R W, Breda, M, Candy, I, Collins, M J, Coope, G R., Durbridge, P, Field, M H, Lee, J R, Lister, A M, Mutch, R., Penkman, K E H, Preece, R C, Rose, J, Stringer, C B, Symmons, R, Whittaker, J E, Wymer, J J and Stuart, A J 2005 'The earliest record of human activity in northern Europe'. *Nature* **438**, 1008–12

Parfitt, S A, Ashton, N M, Lewis, S G, Abel, S G, Coope, G R, Field, M H, Gale, R, Hoare, P G, Larkin, N R, Lewis, M D, Karloukovsk, V, Maher, B A,Peglar, S M, Preece, R C, Whittaker, J E and Stringer, C B 2010 'Early Pleistocene human occupation at the edge of the boreal zone in northwest Europe'. *Nature* **466**, 229–33

Parkes, C 2000 *Fowey Estuary Historic Audit*. Truro: Cornwall Archaeological Unit

Peeters, H, Murphy, P and Flemming, N C (eds) 2009 *North Sea Prehistory Research and Management Framework (NSPRMF) 2009*. Amersfoort: Rijksdienst voor het Cultureel Erfgoed/English Heritage

Petts, D and Gerrard, C 2006 *Shared Visions: The North-East Regional Research Framework for the Historic Environment*. Durham: Durham County Council

Pitt, Sir M 2008 *Lessons Learnt from the 2007 Floods* archive. cabinetoffice.gov.uk/pittreview/thepittreview.html

Ransley, J and Sturt, F 2013 *People and the Sea: a Maritime Archaeological Research Agenda for England*. CBA Research Report 171. York: Council for British Archaeology

Ratcliffe, J 1997 *Fal Estuary Historic Audit*. Truro: Cornwall Archaeological Unit.

Ratcliffe, J and Sharpe, A 1991 *Fieldwork on Scilly, Autumn 1990*. Truro: Cornwall Archaeological Unit

Ratcliffe J and Straker, V 1996 *The Early Environment of Scilly*. Truro: Cornwall Archaeological Unit

Reader, F W 1911 'A Neolithic floor in the bed of the Crouch River and other discoveries near Rayleigh'. *Essex Naturalist* **16**, 249–64

Reid, C 1913 *Submerged Forests*. Cambridge: Cambridge University Press

Reynolds, A 2000 *Helford Estuary Historic Audit*. Truro: Cornwall Archaeological Unit.

Richards, M P 2003, 'Explaining the dietary isotope evidence for the rapid adoption of the Neolithic in Britain', *in* Parker Pearson, M (ed) *Food, Culture and Identity in the Neolithic and Early Bronze Age*. British Archaeological Reports International Series **1117**, 31–6.

Rippon, S 2000 *The Transformation of Coastal Wetlands. Exploitation and Management of Marshland Landscapes in North-West Europe during the Roman and Medieval Periods*. Oxford: Oxford University Press (for the British Academy)

Roberts, M B and Parfitt, S A 2000 *Boxgrove: A Middle Pleistocene Hominid Site at Eartham Quarry, Boxgrove, West Sussex*. English Heritage Archaeological Report **17**. London: English Heritage

Robertson, D and Ames, J 2010. 'Early medieval intertidal fishweirs at Holme Beach, Norfolk'. *Medieval Archaeology* **54**, 329–46

Rowley, T 2006 *The English Landscape in the Twentieth Century*. London: Hambledon Continuum

Scaife, R G 1984 'A history of Flandrian vegetation in the Isles of Scilly: palynological investigation of Higher Moors and Lower Moors peat mines, St Mary's'. *Cornish Studies* II, 33–47

Sear, D A, Bacon, S R, Murdock, A P, Donaghue, G, Baggeley, P, Sera, C and LeBas, T P 2011 'Cartographic, geophysical and diver surveys of the medieval town at Dunwich, Suffolk, England'. *International Journal of Nautical Archaeology* **40 (1)**, 113–32

Selkirk, A 2012 'Folkestone. Roman villa or Iron Age oppidum?'. *Current Archaeology* 262, 22–9

Sheppard, T 1912 *The Lost Towns of the Yorkshire Coast*. London: A Brown and Sons Ltd

Spurrell, F J C 1889 'On the estuary of Thames and its alluvium'. *Proc Geol Soc London* **11**, 210–30

Strachan, D 1995 'Problems and potential of coastal reconnaissance in Essex'. *Aerial Archaeology News* **10**, 28–35

Strachan, D 1997 *Blackwater Estuary Management Plan Area Archaeological Project: Report No. 1. Dating of some intertidal fish weirs in the Blackwater Estuary*. Chelmsford: Essex County Council/RCHM England

Strachan, D 1999 *Blackwater Estuary Management Plan Area Archaeological Project: Report No. 2. Blackwater Foreshore Survey 1995–6: revisiting the Hullbridge sites*. Chelmsford: Essex County Council/RCHM England

Sturt, F and Van de Noort, R 2013 'The Neolithic and Early Bronze Age', *in* Ransley, J and Sturt, F 2013, 50–74

Tangye, M 1991 'A seventeenth century fish cellar at Port Godrevy'. Gwithian. *Cornish Archaeology* **30**, 243–52.

Tann, G 2004 *Lincolnshire Coastal Grazing Marsh. Archaeological and Historical Data Collection. Report for Lincolnshire Wildlife Trust*. Lindsay Archaeological Services Report No. 770. Lincoln: LAS

Timpany, S 2009 *Geoarchaeological Regional Review of Marine Deposits along the Coastline of Southern England*. Research Department Report 004/2009. Swindon: English Heritage

Tomalin, D, Loader, R G and Scaife, R G (eds) 2012 *Coastal Archaeology in a Dynamic Environment. A Solent Case Study*. British Archaeological Reports British Series **568**. Oxford: BAR

Toms, P 2011 *Seabed Prehistory Area 240. Optical Dating of Submarine Cores*. Research Department Reports Series No. 81-2011. Swindon: English Heritage

Truscoe, K 2007 *Rapid Coastal Zone Assessment for the Severn Estuary. Assessment of Environment Agency Lidar Data*. Gloucestershire County Council, Somerset County Council and English Heritage. Swindon: English Heritage

Van de Noort, R 2004 *The Humber Wetlands. The Archaeology of a Dynamic Landscape*. Macclesfield: Windgather Press

Van de Noort, R 2011 *North Sea Archaeologies: a Maritime Biography, 10,000 BC to AD 1500*. Oxford: Oxford University Press

Waddington, C 2007 *Mesolithic Settlement in the North Sea Basin: a case study from Howick, North-East England*. Oxford: Oxbow Books

Wallis, S and Waughman, M 1998 *Archaeology and Landscape in the Lower Blackwater Valley*. East Anglian Archaeology **82**. Chelmsford: Essex County Council

Walsh, M 2013 'Roman, *c* 43 AD to 400', *in* Ransley, J and Sturt, F 2013, 93–112

Warren, S H, Pigott, S, Clark, J G D, Burkitt, M C, Godwin, H and Godwin, M E 1936 'Archaeology of the submerged landscape of the Essex coast'. *Proceedings of the Prehistoric Society* **2.2**, 178–210

Waughman, M 2005 *Archaeology and Environment of Submerged Landscapes in Hartlepool Bay, England*. Tees Archaeology Monograph No. **2**. Hartlepool: Tees Archaeology

Webster, C J 2008 *The Archaeology of South West England. Resource Assessment and Research Agenda*. Taunton: Somerset County Council somerset.gov.uk/archives/hes/downloads/swarf_15.pdf

Wenban-Smith, F F 2003 *Priory Bay Lower Palaeolithic Site, Isle of Wight. Field Evaluation, Final Report*. Unpublished report

Weninger, B, Schulting, R, Bradtmoeller, Clare, L, Collard, M, Edinborough, K, Hilpert, J, Joris, O, Neikus, M, Rohling, E J and Wagner, B 2008 'The catastrophic final flooding of Doggerland by the Storegga Slide tsunami'. *Documenta Praehistorica* **XXXV**, 1–24

Wessex Archaeology 2005a *Artefacts from the Seabed. ALSF: Marine Aggregates and the Historic Environment Year 1 report*. Ref 51541. Salisbury: Wessex Archaeology

Wessex Archaeology 2005b *Marine Aggregate Industry Protocol for the Reporting of Finds of Archaeological Interest*. Salisbury: Wessex Archaeology

Wessex Archaeology 2007 *Seabed Prehistory: Gauging the Effects of Marine Aggregate Extraction. Round 2 Final Report. Volume IV: Great Yarmouth*. Ref 57422.13. Salisbury: Wessex Archaeology

Wessex Archaeology 2010. *Seabed Prehistory: Site Evaluation Techniques (Area 240). Palaeo-environmental Assessment. Interim Report*. Ref 70754.01. Salisbury: Wessex Archaeology

Wessex Archaeology 2011 *Seabed Prehistory: Site Evaluation Techniques (Area 240). Synthesis*. Ref 70754.04. Salisbury: Wessex Archaeology

Westley, K and Bailey, G 2013 'The Palaeolithic', *in* Ransley, J and Sturt, F 2013, 10–29

Wetland Vision Project 2008 *A 50-Year Vision for Wetlands. England's Wetland Landscape: Securing a Future for Nature, People and the Historic Environment*. Sandy: English Heritage/Environment Agency/Natural England/RSPB/The Wildlife Trusts

Wilkinson, T J and Murphy, P L 1995 *Archaeology of the Essex Coast, Volume I: the Hullbridge Survey*. East Anglian Archaeology **71**. Archaeology Section, Essex County Council. ECC: Chelmsford

Wilkinson T J, Murphy, P L, Brown, N and Heppell, E 2012 *The Archaeology of the Essex Coast Volume II. Excavations at the Prehistoric Site of The Stumble*. East Anglian Archaeology **144**. Essex County Council: Chelmsford

Williams, J and Brown, N 1999 *An Archaeological Research Framework for the Greater Thames Estuary*. Chelmsford: Essex County Council, Kent CC, GLAS and TEP.

Williamson, T 2005 *Sandlands. The Suffolk Coast and Heaths*. Macclesfield: Windgather Press.

Woodcock, A 2003 'The archaeological implications of coastal change in Sussex', *in* Rudling, D (ed) *The Archaeology of Sussex to AD 2000*, 1–16. Brighton: University of Sussex

Wymer, J 1999 *The Lower Palaeolithic Occupation of Britain*. Salisbury: Wessex Archaeology/English Heritage

Wymer, J. 2005 'Occupation before the last glaciation: the Palaeolithic', *in* Ashwin, T and Davidson, A *An Historical Atlas of Norfolk* (3rd edn), 13–14. Norwich: Norfolk Museums Service

Wymer, J J and Robins, P A 1994. 'A Long Blade industry beneath Boreal peat at Titchwell, Norfolk'. *Norfolk Archaeology* **XLII (1)**, 13–37

Yates, D 2007 *Land, Power and Prestige: Bronze Age Field Systems in Southern England*. Oxford: Oxbow Books

INDEX

Page numbers in **bold** refer to illustrations. Place and site names have been indexed only where there is an illustration or substantial discussion in the text. In the Appendix (pages 152–156) only area headings have been indexed.